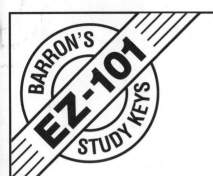

Lawrence S. Leff
Assistant Principal
Mathematics Supervisor
Franklin D. Roosevelt
High School
Brooklyn, New York

College Algebra

All inquiries should be addressed to:
Barron's Educational Series, Inc.
250 Wireless Boulevard
Hauppauge, New York 11788

Library of Congress Catalog Card No. 94-26836

International Standard Book No. 0-8120-1940-7

Library of Congress Cataloging-in-Publication Data
Leff, Lawrence S.
  College algebra / Lawrence S. Leff.
    p. cm. — (Barron's EZ 101 study keys)
  Includes index.
  ISBN 0-8120-1940-7
  1. Algebra. I. Title. II. Series.
QA152.2.L44   1995
512.9—dc20
                                   94-26836
                                          CIP

PRINTED IN THE UNITED STATES OF AMERICA
4567 5500 987654321

CONTENTS

Theme 1 BASIC LAWS AND
OPERATIONS OF ALGEBRA

College Algebra reviews fundamental laws and operations of algebra while introducing more advanced algebraic concepts and methods. This Theme reviews some basic terminology and notation, as well as some elementary algebraic operations.

Key 1 Sets, numbers, and variables

OVERVIEW *Algebra is generalized arithmetic in which letters, called **variables**, are used to represent numbers.*

Set: A **set** is a collection of objects, called **elements**, that are listed within braces, { }. If set *S* contains the first three letters of the alphabet, then *S* = {*a, b, c*}. Since the order of the elements within the braces does not matter, {*b, a, c*} = {*a, b, c*}.

Empty set: A set having no elements is the empty set. If *S* represents the set of elephants that can fly, then *S* is the empty set so *S* = { }.

Null set: The empty set is sometimes called the null set. The null set is denoted by the symbol ∅.

Subsets: If each element of set *B* is also an element of set *A*, then set *B* is a **subset** of set *A*. If *B* = {2, 4} and *A* = {1, **2**, 3, **4**, 5}, then set *B* is a subset of set *A*.

Set of natural (counting) numbers: The set {1, 2, 3 . . . } is the set of natural numbers. The trailing three dots inside the braces indicate that the pattern continues without ending.

Set of whole numbers: The set {0, 1, 2, 3, . . .} is the set of whole numbers.

Set of integers: The set {. . . −3, −2, −1, 0, 1, 2, 3, . . .} is the set of integers.

Set of rational numbers: The set of rational numbers is the set of all numbers that can be written as fractions in which their numerators and denominators are integers. Any integer can be put into this form by writing it as a fraction with a denominator of 1. For example, 3 can be written in fractional form as $\frac{3}{1}$. Thus, the sets of natural numbers, whole numbers, and integers are subsets of the set of rational numbers.

Rational numbers as decimal numbers: A rational number can be written as a nonending repeating decimal number in which a set of one or more digits endlessly repeat, as illustrated below:

$$\frac{1}{4} = 0.25000\ldots, \frac{2}{3} = 0.66666\ldots, \text{ and } \frac{3}{11} = 0.272727\ldots$$

Also, every repeating decimal represents a rational number.

Set of irrational numbers: The set of irrational numbers contains all numbers that cannot be expressed as the quotient of two integers, such as $\sqrt{2}$ and π.

Irrational numbers as decimal numbers: Any nonending decimal number in which there is no repeating pattern of the same sequence of digits is irrational. Thus, 1.4142135. . . (= $\sqrt{2}$) and

0.13133133313333. . . are irrational numbers. Rational numbers have exact decimal equivalents, but irrational numbers do not.

Set of real numbers: The set of real numbers combines the set of rational numbers with the set of irrational numbers.

Real number line: A pictorial representation of the set of real numbers that resembles a "ruler-like" line on which an arbitrary point, called the **origin**, is labeled 0. Positive numbers lie in ascending order to the right of the origin, negative numbers appear in descending order to the left of the origin. As shown in Figure 1.1, each positive number and its negative counterpart (opposite) are located at points that are the same distance from the origin.

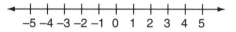

Figure 1.1 The Real Number Line

One-to-one correspondence: The set of real numbers and the set of points on a line can be paired such that each real number corresponds to exactly one point on the line and each point on the line corresponds to exactly one real number. The real number that is paired with a point is called the **coordinate** of that point.

Variable: Often represented by letters from the alphabet such as x and y, a variable is a symbol that serves as a placeholder for an unknown member of a given set, called the **replacement set** or the **domain** of the variable. For example, suppose an unknown integer is increased by 8. We can represent this sum by calling the unknown number x and writing $x + 8$ where the set $\{. . . -3, -2, -1, 0, 1, 2, 3, . . .\}$ gives the possible replacements for x.

Constant. Unlike a variable, a constant has a fixed value. In the sum $x + 8$, 8 is a constant.

Generalizing in algebra: Mathematical laws and properties that apply to all the members of a set of numbers can be concisely stated using variables. For example, if a and b represent real numbers, then "$a + b = b + a$" states that the order in which *any* two real numbers are added does not matter.

Representing multiplication: The result of multiplying two or more quantities together is called a **product**. Each of the different quantities that are being multiplied to form a product is called a **factor** of the product. When writing products we try to avoid using the familiar arithmetic symbol for multiplication, x, since it may be confused with the variable x. The product of real numbers a and b may be represented in any of the following equivalent ways: by writing a and b consecutively, as in ab; or by writing a raised dot between a and b, as in $a \cdot b$; or by using parentheses, as in $(a)(b)$, $a(b)$, or $(a)b$.

Key 2 Equality and inequality relations

OVERVIEW *Special symbols are used to compare one real number to another. Unless otherwise indicated, letters used as variables always represent real numbers.*

Properties of equality: The following properties are assumed to be true:
- Reflexive property: $a = a$.
- Symmetric property: If $a = b$, then $b = a$.
- Transitive property: If $a = b$ and $b = c$, then $a = c$.
- Substitution property: If $a = b$, then in any statement in which a or b appears, one may replace the other.

Order on the number line: Real number a *is greater than* real number b, denoted by $a > b$, if and only if point a lies to the right of point b on the number line. As shown in Figure 1.1, –2 lies to the right of –5 on the number line so $-2 > -5$. Equivalently, –5 *is less than* –2 which is written as $-5 < -2$.

Transitive property of inequality: If $a < b$ and $b < c$, then $a < c$. For example, if $2 < 3$ and $3 < 4$, then $2 < 4$.

Trichotomy property: For any two real numbers exactly one of the following three statements is true:
1. $a < b$ (a is less than b)
2. $a = b$ (a is equal to b)
3. $a > b$ (a is greater than b)

The Symbols $\neq$, $\geq$, and $\leq$: If a and b are not equal, we can write $a \neq b$. We can indicate that a may be equal to *or* greater than b by writing $a \geq b$. The notation $a \leq b$ means a is less than *or* equal to b.

Combining inequality statements: If x is between a and b (with $a < b$) then $x > a$ and $x < b$. The two inequalities can be replaced by the single inequality, $a < x < b$, which means that x is simultaneously greater than a and less than b, as shown in the accompanying diagram.

The absolute value of a: Denoted by $|a|$ and defined such that

$$|a| = a, \text{ if } a \geq 0; \text{ and } |a| = -a, \text{ if } a < 0$$

Thus, $|+4| = 4$ and $|-4| = -(-4) = 4$.

Key 3   Field properties of real numbers

OVERVIEW   *In order to be able to work with real numbers, we need to make some basic assumptions about how they behave. These assumptions are called the **field properties** of real numbers.*

Binary operation: A binary operation, such as addition, works on exactly two numbers of a set at a time. If the outcome of the operation is also a member of the same set, then the set is said to be **closed** under this operation. The set of counting numbers is closed under addition, but not closed under subtraction since the difference between two counting numbers may not be a counting number, as in $2 - 3 = -1$.

Field properties of real numbers: The set of real numbers forms a field since we assume that each of the six field properties listed in Table 1.1 are true for real numbers a, b, and c. A subset of the real numbers is not necessarily a field. For example, the set of integers is not a field since it lacks field property F_5. Although each integer has an additive inverse, each integer does not have a multiplicative inverse. For example, the multiplicative inverse (reciprocal) of 2 is $\frac{1}{2}$ which is *not* a member of the set of integers.

Table 1.1   Field Properties for the Set of Real Numbers

Field Property	Addition	Multiplication
F_1: Closure	$a + b$ is a real number	$a \cdot b$ is a real number
F_2: Commutative	$a + b = b + a$	$a\,b = b\,a$
F_3: Associative	$(a + b) + c = a + (b + c)$	$(a\,b)c = a(b\,c)$
F_4: Identity	0 is the real number such that $a + 0 = 0 + a = 0$	1 is the real number such that $a \cdot 1 = 1 \cdot a = 0$
F_5: Inverse	For each real number a, $-a$ exists such that $a + (-a) = (-a) + a = 0$	For each real number a except 0, $\frac{1}{a}$ exists such that $a\left(\dfrac{1}{a}\right) = \left(\dfrac{1}{a}\right)a = 1$
F_6: Distributive	$a(b + c) = a\,b + a\,c$ and $(b + c)\,a = b\,a + ca$	

Definitions of subtraction and division: For any real numbers a and b,

- The **difference** $a - b$ is defined in terms of addition as $a + (-b)$. For example, $3 - 8 = (3) + (-8) = -5$.

- The **quotient** $a \div b$ is defined in terms of multiplication as $a\left(\dfrac{1}{b}\right)$, where a is called the **dividend**, b is called the **divisor**.

  We assume $b \neq 0$. If dividing by 0 were allowed, then $3 \div 0$ would be equal to some number n. Since $3 \div 0 = n$ implies $3 = n \cdot 0$ and $n \cdot 0 = 0$, we obtain the contradictory statement, $3 = 0$. To avoid this type of contradiction, *division by zero is not defined.* Whenever a variable appears in a denominator of a fraction we assume that it cannot have a value that makes the denominator 0. Thus, in $\dfrac{3}{x-2}$, x cannot be equal to 2.

Taking opposites: A negative sign that precedes a parenthesized expression may be interpreted as "take the opposite of whatever is inside of the parentheses." Thus, $-(-5) = 5$. Sometimes it is helpful to replace the negative sign by -1 and to multiply, using the distributive property, if necessary. For example,

$$
\begin{aligned}
-(2x - 3) &= -1\,(2x - 3) \\
&= (-1)\,2x + (-1)(-3) \\
&= \quad -2x + 3
\end{aligned}
$$

Key 4   Interval notation

OVERVIEW   *A continuous portion of the real number line corresponds to an **interval**. The numbers that belong to an interval can be indicated by writing the left and right boundary points of the interval as an ordered pair.*

Bounded intervals: If $a < b$, then the set of all real numbers from a to b forms a bounded interval with **endpoints** a and b. An interval may or may not include its endpoints. An **open interval** does not include its endpoints, and a **closed interval** includes both of its endpoints. A **half-open interval** (or half-closed interval) includes one endpoint, but not the other. The four possibilities are summarized in Table 1.2. On a graph, a darkened circle on an endpoint means the interval contains that endpoint. An open circle around an endpoint means that endpoint is not included.

Table 1.2   Intervals from a to b where $a < b$

Type of Interval	Inequality Notation	Interval Notation	Graph of Interval
Open	$a < x < b$	(a, b)	
Half-Open	$a \leq x < b$	$[a, b)$	
Half-Open	$a < x \leq b$	$(a, b]$	
Closed	$a \leq x \leq b$	$[a, b]$	

Unbounded intervals: Intervals that extend without bound on one or both sides can be represented by the symbol for "infinity" (unbounded), ∞. The symbol ∞ does not represent a real number. Thus,

- $(a, +\infty)$ means $x > a$
- $[a, +\infty)$ means $x \geq a$
- $(-\infty, a)$ means $x < a$
- $(-\infty, a]$ means $x \leq a$

The entire real number line corresponds to the interval $(-\infty, +\infty)$.

Key 5   Integer exponents

OVERVIEW   *Repeated multiplication of the same quantity may be written in a more compact form using an exponent that tells the number of times the quantity appears as a factor in a product.*

x^n: This expression denotes the product in which x appears as a factor n times. Thus, $2^4 = 2 \cdot 2 \cdot 2 \cdot 2 = 16$. The symbol x^n is read as "x raised to the nth power" where x is called the **base** and n is the **exponent** of the base. When a variable appears without an exponent, the exponent is understood to be one. Thus, $x = x^1$.

Laws of exponents: Powers of the same nonzero base can be multiplied by adding their exponents, or can be divided by subtracting their exponents. These rules are summarized in Table 1.3 where a and b are not 0, and x, y, and z are positive integers.

Table 1.3   Laws of Exponents

Law	Rule	Example
Multiplication	$a^x \cdot a^y = a^{x+y}$ $a^x \cdot b^x = (ab)^x$	$n^5 \cdot n^2 = n^7$ $m^3 \cdot n^3 = (mn)^3$
Division	$a^x \div a^y = a^{x-y} \, (x > y)$ $a^x \div a^y = \dfrac{1}{a^{y-x}} \, (y > x)$ $a^x \div b^x = \left(\dfrac{a}{b}\right)^x$	$n^5 \div n^2 = n^3$ $n^3 \div n^7 = \dfrac{1}{n^4}$ $m^3 \div n^3 = \left(\dfrac{m}{n}\right)^3$
Power of a Power	$(a^x)^y = a^{xy}$	$(n^5)^2 = n^{10}$
Power of a Product	$(a^x b^y)^z = a^{xz} b^{yz}$	$(mn^4)^3 = m^3 n^{12}$

Zero exponent: Any nonzero quantity raised to the zero power is 1. Thus, $x^0 = 1$ provided $x \neq 0$. The expression 0^0 is undefined.

Negative integer exponents: If n is a positive integer and x, $y \neq 0$, then

$$x^{-n} = \frac{1}{x^n}, \; \frac{1}{x^{-n}} = x^n$$

and

$$\left(\frac{x}{y}\right)^{-n} = \left(\frac{y}{x}\right)^{n} = \frac{y^n}{x^n}$$

Examples: 1. $2^{-3} = \frac{1}{2^3} = \frac{1}{8}$

2. $\frac{1}{4^{-2}} = 4^2 = 16$

Scientific notation: Scientific notation is used to express a positive number as a number greater than or equal to 1 and less than 10 multiplied by an integer power of 10.

- To write a number greater than 10 in scientific notation:
  1. Move the decimal point k places to the left so that the number is now between 1 and 10.
  2. Multiply by 10^k.

  Thus,

  $1\,2\,0\,0\,0\,0\,0.$

  $k = 6$

  1.2×10^6

  $1{,}200{,}000 = 1.2 \times 10^6$

- To write a positive number less than 1 in scientific notation:
  1. Move the decimal point k places to the right so that the number is now between 1 and 10.
  2. Multiply by 10^{-k}.

  Thus,

  $0.0\,0\,0\,0\,3\,0\,8$

  $k = 5$

  3.08×10^{-5}

  $0.0000308 = 3.08 \times 10^{-5}$

- To write a positive number between 1 and 10 in scientific notation, multiply the number by 10^0. In scientific notation, $6.14 = 6.14 \times 10^0$.

Order of operations: Arithmetic operations are not necessarily performed in the order in which they are encountered. Instead,

1. First, evaluate powers.
2. Next, perform multiplications and divisions, working from left to right.
3. After that, perform additions and subtractions, again working from left to right.

If an expression contains parentheses or brackets, then the expression inside the grouping symbol is evaluated first.

Example 1.
$$5 + 18 \div (4 - 1)^2 = 5 + 18 \div 3^2$$
$$= 5 + 18 \div 9$$
$$= 5 + 2$$
$$= 7$$

Example 2. If $x = -2$ and $y = 5$, then

$$-y^2 + 3x^2 = -5^2 + 3(-2)^2$$
$$= -25 + 3(4)$$
$$= -25 + 12$$
$$= -13$$

Key 6 Polynomials

OVERVIEW *Variables and numbers may be multiplied and their products combined to produce expressions called **polynomials**. Since polynomials represent real numbers, they can be added, subtracted, multiplied, and divided using the properties of real numbers.*

Monomial: A monomial is a number, a variable, or the indicated *product* of a number and one or more variables with positive exponents like 3, $-2xy$, and $5a^3b^2$. The numerical factor of a monomial is called its **coefficient**. The numerical coefficient of $-3ab$ is -3. The numerical coefficient of x^2 is 1 since $x^2 = 1 \cdot x^2$.

Like monomials: Monomials that have the same variables raised to the same powers, as in $3a^2b$, $-5a^2b$, and a^2b, are like monomials. They can be added or subtracted by combining their numerical coefficients.

Example 1. $3x^2 + 4x^2 = (3 + 4) x^2 = 7x^2$

Example 2. $7mn - mn = (7 - 1) mn = 6mn$

Polynomial: A polynomial is a monomial or the sum of two or more monomials. Each monomial in the sum is called a **term** of the polynomial. A polynomial having two unlike terms, like $2x + 3y$, is called a **binomial**. Since the polynomial $x^2 - 3x + 7$ has three unlike terms, it is called a **trinomial.**

The degree of a polynomial: The largest number obtained by adding the exponents of any term of a polynomial is its degree. The degree of $4xy + 7$ is 2 since $4xy = 4x^1y^1$ and $1 + 1 = 2$. The degree of $2x^3 - 5x^2 + 4x - 7$ is 3 and the degree of $x^2y + x^2y^3 + y^4$ is 5 since the sum of the exponents in the middle term is $2 + 3 = 5$. The degree of a polynomial is 0 when its only term is a nonzero constant.

Standard form: A polynomial in one variable is in **standard form** when the powers of the variable decrease in value as the terms are read from left to right. The polynomial $8x^3 + 6x^4 + 5 - 2x$ is not in standard form until its terms are rearranged so that the resulting polynomial is $6x^4 + 8x^3 - 2x + 5$.

Operations with polynomials:

- To add polynomials, collect and then add their like terms.

  Example:
  $$(4x^2 + 3y - 8) + (5x^2 - 2y + 7) = (4x^2 + 5x^2) + (3y - 2y) + (-8 + 7)$$
  $$= 9x^2 + y - 1$$

- To subtract polynomials take the opposite of each term of the polynomial that is being subtracted, then add the two polynomials.

  Example:
  $$(3a^2 - 7a + 5) - (2a^2 - 4a - 1) = (3a^2 - 7a + 5) + (-2a^2 + 4a + 1)$$
  $$= (3a^2 - 2a^2) + (-7a + 4a) + (5 + 1)$$
  $$= a^2 - 3a + 6$$

- To multiply monomials, multiply their coefficients and then multiply the variable factors by *adding* the exponents of like variables.

  Example:
  $$(-2a^2b)(4a^3b^2) = (-2)(4)(a^2\ a^3)(b\ b^2)$$
  Note that $b = b^1$:
  $$= -8\ (a^{2+3})\ (b^{1+2})$$
  $$= -8a^5b^3$$

- To divide monomials, divide their numerical coefficients and then divide the variable factors by *subtracting* the exponents of like variables.

  Example:
  $$\frac{14a^5b^4c}{21ab^3c^2} = \frac{2 \cdot 6}{3 \cdot 7}(a^{5-1})(b^{4-3})(c^{1-2})$$
  $$= \frac{2}{3}a^4b^1c^{-1} = \frac{2}{3}\frac{a^4b}{c}$$

- To multiply a polynomial by a monomial, multiply each term of the polynomial by the monomial, and then add the resulting products.

  Example:
  $$x^2\ (3x^2 - 5x + 1) = (x^2)(3x^2) + x^2\ (-5x) + x^2(1)$$
  $$= 3x^4 - 5x^3 + x^2$$

- To multiply a binomial by a polynomial, multiply each term of the polynomial by the binomial and then simplify using the distributive law.

  Example:
  $$(2x - 3)(x^2 - 5x + 4) = (2x - 3)x^2 + (2x - 3)(-5x) + (2x - 3)(4)$$
  $$= (2x^3 - 3x^2) + (-10x^2 + 15x) + (8x - 12)$$
  $$= 2x^3 - 13x^2 + 23x - 12$$

- To divide a polynomial by a monomial, divide each term of the polynomial by the monomial and then add the quotients.

  Example:
  $$\frac{12x^3 - 3x^2 + 6x}{3x} = \frac{12x^3}{3x} - \frac{3x^2}{3x} + \frac{6x}{3x}$$
  $$= 4x^2 - x + 2$$

FOIL method for multiplying binomials: This method offers a short-cut for multiplying two binomials. The product of the binomials $(a + b)$ and $(c + d)$ is equal to the sum of the products of their *F*irst terms $(a \cdot c)$, *O*uter terms $(a \cdot d)$, *I*nner terms $(b \cdot c)$, and *L*ast terms $(b \cdot d)$.

Example:

	F	**O**	**I**	**L**
$(3x - 7)(2x + 5) =$	$(3x)(2x) +$	$[(3x)(5) +$	$(-7)(2x)] +$	$(-7)(5)$
$=$	$6x^2 +$	$[15x - 14x]$	$+$	(-35)
$=$	$6x^2 +$	x		-35

Squaring binomials: The square of a binomial is a trinomial having one of the following forms:

- $(a + b)^2 = a^2 + 2ab + b^2$
- $(a - b)^2 = a^2 - 2ab - b^2$

  Example:
  You can expand $(x + 2y)^2$ by using the formula
  $$(a + b)^2 = a^2 + 2ab + b^2$$
  where $a = x$ and $b = 2y$. Thus,
  $$(x + 2y)^2 = x^2 + 2(x)(2y) + (2y)^2$$
  $$= x^2 + 4xy + 4y^2$$

Long division of polynomials: When the divisor of a polynomial is not a monomial, the quotient is obtained using a procedure similar to the one used to divide whole numbers in arithmetic. For example, proceed as follows to divide:

$$6x^3 - x^2 - 8x - 7 \text{ by } 2x - 3$$

Step 1 Divide the first term of the dividend by the first term of the divisor to obtain the first term of the quotient. The result of dividing $6x^3$ by $2x$ is $3x^2$.

$$
\begin{array}{r}
3x^2 \\
2x-3 \overline{)6x^3 - x^2 - 8x - 7}
\end{array}
$$

Step 2 Multiply the entire divisor by the first term of the quotient and then subtract this product from the dividend. The difference (remainder) is $8x^2 - 8x$.

$$
\begin{array}{r}
3x^2 \\
2x-3 \overline{)6x^3 - x^2 - 8x - 7} \\
\underline{-6x^3 - 9x^2 } \\
8x^2 + 8x
\end{array}
$$

Step 3 Using this remainder as the new dividend, repeat Steps 1 and 2 until a constant remainder is obtained.

$$
\begin{array}{r}
3x^2 + 4x + 2 \\
2x-3 \overline{)6x^3 - x^2 - 8x - 7} \\
\underline{-6x^3 - 9x^2 } \\
8x^2 - 8x \\
\underline{-8x^2 - 12x } \\
4x - 7 \\
\underline{-4x - 6 } \\
-1
\end{array}
$$

The quotient is $3x^2 + 4x + 2$ with a remainder of -1.
Thus,

$$\frac{6x^3 - x^2 - 8x - 7}{2x - 3} = 3x^2 + 4x + 2 + \frac{-1}{2x - 3}$$

Step 4 Check by multiplying the quotient by the divisor and then adding any remainder. The check is left for you.

Key 7   Factoring Polynomials

OVERVIEW   *Factoring a polynomial means writing the polynomial as the product of two or more lower degree polynomials each of which is called a **factor** of the original polynomial.*

Factoring over the set of integers: We usually factor a polynomial so that if the original polynomial has integer coefficients, so do each of its factors.

Removing the greatest common monomial factor: If all the terms of a polynomial contain a common monomial factor, then it can be removed by using the reverse of the distributive property.

Example 1:   The greatest common factor (GCF) of the terms of $3x^4 + 12x$ is $3x$ since 3 and x are the greatest numerical and variable factors common to each term of the polynomial. Thus, $3x^4 + 12x = 3x (x^3 + 4)$. You can check that the factorization of $3x^4 + 12x$ is correct by multiplying the two factors together and verifying the product is the original polynomial.

Example 2:   The greatest number that evenly divides the numerical coefficients in $24a^3b - 30a^2 b^5$ is 6. The greatest powers of a and b that are common to both terms of the polynomial are a^2 and b^1. Hence, the GCF of the two terms of the polynomial is $6a^2b$. The corresponding factor of the polynomial can be obtained by dividing the polynomial by $6a^2 b$ which gives $4a - 5b^4$. Thus, $24a^3b - 30a^2 b^5 = 6a^2b (4a - 5b^4)$.

Factoring by grouping: Some polynomials can be factored by grouping its terms so that the reverse of the distributive property can be applied in successive stages. For example, the polynomial $xz + yz + xw + yw$ can be factored by first grouping the first two terms and the last two terms together, as in $z (x + y) + w (x + y)$, and then factoring out $(x + y)$.
Thus,

$$xz + yz + xw + yw = z (x + y) + w (x + y)$$
$$= (x + y) (z + w)$$

Example:

$$8mx + 4px - 6m - 3p = (8mx + 4px) + (- 6m - 3p)$$
$$= 4x (2m + p) - 3 (2m + p)$$
$$= (2m + p) (4x - 3)$$

Special factoring patterns: Sometimes a polynomial can be factored by recognizing that it has the same form as a special polynomial whose factorization is already known. See Table 1.4.

Table 1.4 Special Factoring Patterns

Factorization Formula	Example
Difference of Two Squares: $a^2 - b^2 = (a - b)(a + b)$	Factor $4x^2 - 9$. Let $a = 2x$ and $b = 3$. Then, $4x^2 - 9 = (2x - 3)(2x + 3)$
Sum of Two Cubes: $a^3 + b^3 = (a + b)(a^2 - ab + b^2)$	Factor $x^3 + 8y^3$. Let $a = x$ and $b = 2y$. Then, $x^3 + 8y^3 = (x + 2y)(x^2 - 2xy + 4y^2)$
Difference of Two Cubes: $a^3 - b^3 = (a - b)(a^2 + ab + b^2)$	Factor $27x^3 - y^3$. Let $a = 3x$ and $b = y$. Then, $27x^3 - y^3 = (3x - y)(x^2 + 3xy + y^2)$

Factoring $x^2 + bx + c$: If $x^2 + bx + c$ can be factored as the product of two binomials, then $x^2 + bx + c = (x + p)(x + q)$ where the numbers p and q are chosen so that their sum is b and their product is c.

- If $c > 0$, then p and q must have the same sign. For example, $x^2 + 7x + 12 = (x + 3)(x + 4)$ and $x^2 - 7x + 12 = (x - 3)(x - 4)$

- If $c < 0$, then p and q must have opposite signs. For example, $x^2 + x - 12 = (x + 4)(x - 3)$ and $x^2 - x - 12 = (x - 4)(x + 3)$

Factoring $ax^2 + bx + c$ ($a \neq 0$): If $ax^2 + bx + c$ can be factored as the product of two binomials, then

$$ax^2 + bx + c = (rx + p)(sx + q)$$

such that

$$rs = a, \quad ps + rq = b, \quad \text{and} \quad pq = c .$$

Not all quadratic trinomials can be factored over the set of integers.

KEY EXAMPLE

Factor: (a) $x^2 + 3x - 10$ (b) $4x^2 + 5x - 6$

Solution: (a) To factor $x^2 + 3x - 10$, find a pair of numbers, p and q, whose sum is 3 and whose product is -10. Letting $p = 5$ and $q = -2$, gives the correct factorization: $x^2 + 3x - 10 = (x + 5)(x - 2)$.

(b) To factor $4x^2 + 5x - 6$, find two numbers, r and s, such that their product is 4 and another pair of numbers, p and q, whose product is -6. The positive factors of 4 are:

$$2, 2 \text{ and } 4, 1.$$

Thus,

$$4x^2 + 5x - 6 = (2x + p)(2x + q)$$

or

$$4x^2 + 5x - 6 = (4x + p)(x + q)$$

The factors of -6 are: $-6, 1, 6, -1, 3, -2,$ and $-3, 2$. By trial and success, we find that letting $r = 4$, $s = 1$, $p = -3$, and $q = 2$ gives the correct factorization:

$$4x^2 + 5x - 6 = (4x - 3)(x + 2)$$

Factoring completely: A **prime polynomial** has no factors other than itself and 1 (or -1). A polynomial is **factored competely** when each of its factors are prime polynomials so that they cannot be factored further. To factor completely, it may be necessary to use more than one factoring technique. Typically, we begin by removing a common monomial factor, if possible, and then factoring further, again if possible.

KEY EXAMPLE

Factor completely:

   (a) $3y^3 - 75y$ (b) $x^{4n} - 81$ (c) $a^3b + 2a^3 - b - 2$

Solution: (a) $3y^3 - 75y = 3y(y^2 - 25)$

Factor $y^2 - 25$: $= 3y(y - 5)(y + 5)$

   (b) $x^{4n} - 81 = (x^{2n} - 9)(x^{2n} + 9)$

Factor $x^{2n} - 9$: $= (x^n - 3)(x^n + 3)(x^{2n} + 9)$

   (c) $a^3b + 2a^3 - b - 2 = a^3(b + 2) - (b + 2)$

   $= (b + 2)(a^3 - 1)$

Factor $a^3 - 1$: $= (b + 2)(a - 1)(a^2 + a + 1)$

Key 8   Radicals and rational exponents

OVERVIEW   *The* k*th root of* x, *denoted by* $\sqrt[k]{x}$ *or* $x^{\frac{1}{k}}$, *is one of* k *identical real numbers whose product is* x.

Square root of *x*:   A square root is one of two identical real numbers whose product is x.   The square root of a nonnegative number x is denoted by $\sqrt{x}$. In $\sqrt{x}$, the symbol $\sqrt{}$ is called a **radical sign** and x is called the **radicand**. For example, $\sqrt{16} = 4$ since $4 \cdot 4 = 16$.

However, $\sqrt{-9}$ does not exist as the product of two identical real numbers can never equal a negative number.

Cube Root: A **cube root** of a number is one of three identical numbers whose product is the number. Thus, $\sqrt[3]{8} = 2$ since $2 \cdot 2 \cdot 2 = 8$ and $\sqrt[3]{-8} = -2$ since $(-2)(-2)(-2) = -8$. The *3* in $\sqrt[3]{-8}$ tells what root of x is to be taken and is called the **index** of the radical. In square root radicals the index is usually omitted. Thus, $\sqrt[2]{64}$ means $\sqrt{64}$.

The *k*th root of *x*:   A real number b is the kth root of x, denoted by $\sqrt[k]{x}$, if and only if b satisfies the equation $b^k = x$. For example, 2 is a fourth root of 16 ($\sqrt[4]{16} = 2$) since $2^4 = 16$ and -4 is a cube root of -64 since $(-4)^3 = -64$.   If $x < 0$ and k is even, then $\sqrt[k]{x}$ is not a real number so the kth root of x does not exist.

Principal *k*th root: The notation $\sqrt[k]{x}$ always refers to the **principal *k*th root of *x*** which is the real valued root, if it exists, that has the same sign as x whenever $x \neq 0$. For example, although the two square roots of 16 are 4 and -4, only 4 is the *principal* square root of 16 since 4 has the same sign as 16. Thus, $\sqrt{16} = 4$.

Definition of $x^{\frac{1}{k}}$: The principal kth root of x provided $x \geq 0$ whenever k is even. Thus,

$$x^{\frac{1}{k}} = \sqrt[k]{x}$$

Examples:

$$36^{\frac{1}{2}} = \sqrt{36} = 6 \text{ and } (-125)^{\frac{1}{3}} = \sqrt[3]{-125} = -5$$

Definition of $x^{n/k}$:   If $\sqrt[k]{x^n}$ is a real number, then

$$x^{n/k} = (\sqrt[k]{x})^n = \sqrt[k]{x^n}$$

where x is any real number, n is an integer, and k is a positive integer.

In evaluating expressions of the form $x^{\frac{n}{k}}$, it is usually easier to first take the kth root of x and then raise the result to the nth power, as in

$$27^{\frac{2}{3}} = (\sqrt[3]{27})^2 = (3)^2 = 9 .$$

Laws for rational exponents: The laws of positive integer exponents given in Table 1.3 also hold for exponents that are rational numbers.

Key 9   Operations with radicals

OVERVIEW   *Radicals may be multiplied and divided, provided that they have the same index. Radicals that have the same index and the same radicand can be added or subtracted.*

Multiplication rule for radicals: $c\sqrt[k]{a}\ d\sqrt[k]{b} = cd\sqrt[k]{ab}$, provided $ab \geq 0$

if k is even. For example, $2\sqrt{5} \cdot 3\sqrt{7} = 6\sqrt{35}$.

Quotient rule for radicals:

$$\frac{c\sqrt[k]{a}}{d\sqrt[k]{b}} = \frac{c}{d}\sqrt[k]{\frac{a}{b}}$$

(when ($b \neq 0$ and $ab \geq 0$ if k is even).

Example:

$$\frac{12\sqrt{21}}{6\sqrt{7}} = \frac{12}{6}\sqrt{\frac{21}{7}} = 2\sqrt{3}$$

Perfect square: Any term that can be expressed as the product of two identical factors is a perfect square. Forty-nine is a perfect square since $49 = 7 \cdot 7$.

Simplifying radicals: To simplify $\sqrt{N}$, write $\sqrt{N} = \sqrt{a} \cdot \sqrt{b}$ where

a is the greatest perfect square factor of N. Then evaluate $\sqrt{a}$.

Example 1. $\sqrt{28}$ $=$ $\sqrt{4} \cdot \sqrt{7}$ $=$ $2\sqrt{7}$
Example 2. $\sqrt{48x^5}$ $=$ $\sqrt{16x^4} \cdot \sqrt{3x}$ $=$ $4x^2\sqrt{3x}$

In Example 2, x^4 is the greatest perfect square factor of x^5 since its exponent is the greatest power contained in x^5 that is divisible by 2, the index of the radical. Radicals with indexes other than 2 can be simplified using a similar procedure.

Example 3. $\sqrt[3]{24y^7}$ $=$ $\sqrt[3]{8y^6} \cdot \sqrt[3]{3y}$ $=$ $2y^2\sqrt[3]{3y}$

In Example 3, 8 is the greatest perfect cube factor of 24 and y^6 is the greatest perfect cube factor of y^7. The factor y^6 is obtained by finding the greatest factor of y^7 whose exponent is evenly divisible by the index of the radical, 3.

Like radicals: Like radicals have the same index and the same radicand. Like radicals can be added or subtracted by combining their coefficients, as in $2\sqrt{5} + 7\sqrt{5} = (2+7)\sqrt{5} = 9\sqrt{5}$. Sometimes unlike radicals can be combined after the radicals have been simplified.

Example:

$$\begin{aligned} 3\sqrt{48} + \sqrt{75} &= 3\sqrt{16}\sqrt{3} + \sqrt{25}\sqrt{3} \\ &= 3 \cdot 4\sqrt{3} + 5\sqrt{3} \\ &= 12\sqrt{3} + 5\sqrt{3} \\ &= 17\sqrt{3} \end{aligned}$$

Multiplying radical expressions: Two radical expressions that each have the form of a binomial can be multiplied using FOIL.

Example:

$$\overset{\text{F} \quad\quad \text{O} \quad\quad \text{I} \quad\quad \text{L}}{(2 + 3\sqrt{5})(1 - \sqrt{7}) = 2 - 2\sqrt{7} + 3\sqrt{5} - 3\sqrt{35}}$$

Conjugate radical expressions: Pairs of radical expressions of the form $(A + B\sqrt{C})$ and $(A - B\sqrt{C})$ are called conjugate radicals. The product of conjugate radicals does not include a radical since, using FOIL,

$$(A + B\sqrt{C})(A - B\sqrt{C}) = A^2 - B^2C$$

Example: The product $(7 + 2\sqrt{3})(7 - 2\sqrt{3})$ can be obtained by letting $A = 7$, $B = 2$, and $C = 3$. Thus,

$$(7 + 2\sqrt{3})(7 - 2\sqrt{3}) = 7^2 - 2^2(3) = 49 - 12 = 37.$$

Rationalizing a denominator: This is the process of changing a fraction with a radical in its denominator into an equivalent fraction whose denominator has no radical. To rationalize a denominator of the form $B\sqrt{C}$, multiply the fraction by "1" in the form of $\dfrac{\sqrt{C}}{\sqrt{C}}$.

Example:

$$\frac{5}{\sqrt{3}} = \frac{5}{\sqrt{3}} \cdot \left(\frac{\sqrt{3}}{\sqrt{3}}\right) = \frac{5\sqrt{3}}{\sqrt{9}} = \frac{5\sqrt{3}}{3}$$

Similarly, to rationalize a denominator of the form $A + B\sqrt{C}$ multiply the numerator and the denominator of the fraction by its conjugate, $A - B\sqrt{C}$.

Example: Rationalize the denominator of $\dfrac{\sqrt{8}}{5-3\sqrt{2}}$ by using the conjugate of the denominator, $5+3\sqrt{2}$, as the rationalizing factor. Thus,

$$\frac{\sqrt{8}}{5-3\sqrt{2}} \cdot \left(\frac{5+3\sqrt{2}}{5+3\sqrt{2}}\right) = \frac{\sqrt{8}(5+3\sqrt{2})}{5^2-3^2(2)}$$

$$= \frac{5\sqrt{8}+3\sqrt{16}}{25-18}$$

$$= \frac{5\sqrt{8}+12}{7}$$

Theme 2 EQUATIONS AND INEQUALITIES

Finding solutions to various types of equations and inequalities is an important aspect of algebra. Solving a linear equation requires isolating the variable on one side of the equation and placing the constant terms on the other side. To do this we need to undo whatever operations are "attached" to the variable by performing their inverses. For example, if $x + 3 = 7$ then we can cancel the effect of *adding* 3 to x by *subtracting* 3 from each side of the equation. This gives $x = 4$. Solving equations that are not linear may also involve performing inverse operations. If $\sqrt{x} = 5$, we solve for x by undoing the square root of x by raising both sides of the equation to the second power. This gives $x = 25$. To solve $x^2 - 3x = 0$, we undo the multiplication that led to $x^2 - 3x$, by factoring the left-side of the equation as $x(x - 3) = 0$ and then setting each factor equal to 0 so that $x = 0$ or $x = 3$.

Key 10 Linear equations

OVERVIEW *An **equation** states that two expressions represent the same number. An equation that contains a variable cannot be judged true or false until the variable is replaced with a number.*

Conditional equation: A conditional equation is one that is true for some values of the variable, but false for other values. The equation $2x - 1 = x + 3$ is true for $x = 4$, and is false for any other replacement for x. Any replacements of the variable that make an equation a true statement are called **solutions** or **roots** of the equation. Thus, 4 is a solution or root of the equation $2x - 1 = x + 3$.

Identity: An identity is an equation that is true for all possible replacements for x. The equation $2(x + 1) = 2x + 2$ is an identity.

Solution set: The collection of all solutions of an equation is the solution set.

Solving an equation: The process of finding the solution set for an equation is referred to as "solving" that equation.

Equivalent equations: These equations have the same solution set. The equations $x + 1 = 4$, $2x = 6$, and $x = 3$ are equivalent equations since they have the same solution, $x = 3$.

Properties of equations: The following properties may be needed when solving equations:

- $\mathbf{P_1}$ Adding or subtracting the same quantity on both sides of the equation produces an equivalent equation. For example, if

$$x + 1 = 3,$$

  then

$$x + 1 - (1) = 4 - (1)$$

  so

$$x = 3.$$

- $\mathbf{P_2}$ Multiplying or dividing both sides of the equation by the same *nonzero* quantity produces an equivalent equation. For example, if

$$3x = 6,$$

  then

$$\frac{3x}{3} = \frac{6}{3}$$

  so

$$x = 2.$$

Linear (first-degree) equation: In a linear equation the exponent of each variable is 1. For example, $2x - 1 = x + 3$ is a linear equation, but the equation $x^2 = 4$ is not linear. A linear equation in one variable, say x, is an equation that can be put into the form $ax + b = 0$, where a and b are constants with $a \neq 0$.

Strategy for solving linear equations: Use the properties of equations to obtain successively simpler equivalent equations until the variable is isolated by itself on one side of the equation and a constant term is by itself on the other side.

KEY EXAMPLE

Solve for x: $\qquad 5x - 8 = 2x + 13$

Solution: Collect variable terms on the left-side of the equation and the constant terms on the right-side of the equation.

By P_1:	$5x - 8 + (8) = 2x + 13 + (8)$
Simplify:	$5x = 2x + 21$
By P_1:	$5x - (2x) = 2x + 21 - (2x)$
Simplify:	$3x = 21$
By P_2:	$\dfrac{3x}{3} = \dfrac{21}{3}$
Simplify:	$x = 7$

KEY EXAMPLE

Solve for x: $\qquad 2(7 - 3x) = 3(x + 1) + 29$

Solution: Remove the parentheses by multiplying each term inside the parentheses by the factor outside the parentheses. Then collect variable terms on the left side of the equation and constants on the right side of the equation.

	$2(7 - 3x) = 3(x + 1) + 29$
Use the Distributive Law:	$2(7) - 2(3x) = 3(x) + 3(1) + 29$
Simplify:	$14 - 6x = 3x + 32$
By P_1:	$14 - 6x - (14) = 3x + 32 - (14)$
Simplify:	$-6x = 3x + 18$

By P_1:	$-6x - (3x) = 3x + 18 - (3x)$
Simplify:	$-9x = 18$
By P_2:	$\dfrac{-9x}{-9} = \dfrac{18}{-9}$
	$x = -2$

Checking roots: To check that a number is a root or solution of an equation, replace the variable by the root in the original equation and verify that the left-hand and the right-hand sides of the original equation represent the same number.

Example: To check that $x = 7$ is a root of $5x - 8 = 2x + 13$, replace x with 7 and then compare the values of the left-hand and the right-hand sides of the equation:

$$5(7) - 8 \overset{?}{=} 2(7) + 13$$

$$35 - 8 \overset{?}{=} 14 + 13$$

$$\checkmark$$

$$27 = 27$$

To solve linear equations with fractions: Clear the equation of its fractions by multiplying each term of the equation by the least common multiple (LCM) of the denominators of the fractional terms.

Example: To solve $\dfrac{x+1}{4} - \dfrac{2}{3} = \dfrac{x-7}{12}$,

write an equivalent equation without fractions by multiplying each term of the equation by 12 which is the least common multiple of 4, 3, and 12. Thus,

$$12\left(\frac{x+1}{4}\right) - 12\left(\frac{2}{3}\right) = 12\left(\frac{x-7}{12}\right)$$

The resulting equation is

$$3(x + 1) - 8 = x - 7$$

whose solution you can verify as

$$x = -1.$$

Proportion: A proportion is an equation having the form $\dfrac{a}{b} = \dfrac{c}{d}$.

Cross-products rule: If

$$\frac{a}{b} = \frac{c}{d}, \text{ then } bc = ad.$$

Example: To solve the proportion

$$\frac{x+5}{4} = \frac{x-2}{3}$$

write the equivalent equation

$$4(x-2) = 3(x+5)$$

whose solution you can verify is

$$x = 23.$$

Literal equation: An equation with more than one letter or variable is a literal equation. Solving a literal equation for a specified letter or variable requires isolating this letter or variable in much the same way that an equation having a single variable is solved.

Example: To solve

$$ay = by + c \text{ for } y,$$

isolate y on the left side of the equation by subtracting by from both sides of the equation:

$$ay - by = by + c - by$$

Factor out y: $\qquad y(a-b) = c$

Divide by the coefficient of y: $\quad \dfrac{y(a-b)}{a-b} = \dfrac{c}{a-b}$

$$y = \frac{c}{a-b} \text{ (provided } a \neq b)$$

Key 11 Linear inequalities

OVERVIEW *Replacing the equal sign in a linear equation with an inequality symbol creates a **linear inequality**. Linear inequalities are solved in much the same way that linear equations are solved.*

Properties of inequality: An inequality can be changed into an equivalent inequality by performing the following operations:

P_3 Add or subtract the same quantity on both sides of an inequality. For example, if $x - 1 < 3$, then adding 1 to each side gives the equivalent inequality, $x < 4$.

P_4 Multiply or divide both sides of an inequality by the same *positive* quantity. For example, if $3x < 6$, then dividing each side by 3 gives the equivalent inequality $x < 2$.

P_5 Multiply or divide both sides of an inequality by the same *negative* quantity and then reverse the direction of the inequality sign. For example, if $-2x > 8$, then dividing each side by -2 and reversing the direction of the inequality from $>$ to $<$ gives the equivalent inequality $x < -4$.

KEY EXAMPLE

Solve and graph the solution set of

$$1 - 2x < x + 13.$$

Solution: Isolate the variable on the left side of the inequality:

By P_3: $\qquad\qquad\qquad 1 - 2x - 1 < x + 13 - 1$

$\qquad\qquad\qquad\qquad\qquad -2x < x + 12$

By P_3: $\qquad\qquad\qquad -x - 2x < x + 12 - x$

$\qquad\qquad\qquad\qquad\qquad -3x < 12$

By P_5 : $\qquad\qquad\qquad \dfrac{-3x}{-3} > \dfrac{12}{-3}$ (direction of inequality is reversed)

$\qquad\qquad\qquad\qquad\qquad x > -4$

The solution is $x > -4$, which when graphed consists of all points greater than, but not equal to, -4.

Compound inequality: A compound inequality combines two inequalities into an extended inequality. Thus, the solution of the compound inequality $3 < 2x - 1 < 5$ must satisfy the two inequalities $3 < 2x - 1$ and $2x - 1 < 5$ simultaneously. Performing the same arithmetic operation with each member of a compound inequality produces an equivalent inequality. Thus, adding 1 to each member of $3 < 2x - 1 < 5$ produces the equivalent inequality $4 < 2x < 6$. Then dividing each member of this inequality by 2 gives $2 < x < 3$.

KEY EXAMPLE

Solve and graph the solution:

$$-1 \le \frac{4 - 7x}{3} < 6$$

Solution: Clear the fraction by multiplying each member of the inequality by 3:

$$(3)(-1) \quad \le \quad (3)\frac{4 - 7x}{3} \quad < \quad (3)(6)$$

$$-3 \quad \le \quad 4 - 7x \quad < \quad 18$$

$$-3 - 4 \quad \le \quad 4 - 7x - 4 \quad < \quad 18 - 4$$

$$-7 \quad \le \quad -7x \quad < \quad 14$$

Divide by -7 and reverse the direction of the inequality:

$$\frac{-7}{-7} \quad \ge \quad \frac{-7x}{-7} \quad > \quad \frac{14}{-7}$$

$$1 \quad \ge \quad x \quad > \quad -2$$

or, equivalently, $\quad -2 \quad < \quad x \quad \le \quad 1$

The solution, $-2 < x \le 1$, represents the interval $(-2, 1)$, the graph of which is shown.

Key 12 Absolute value equations and inequalities

OVERVIEW *Absolute value equations and inequalities are solved by writing equivalent statements that do not involve absolute value. This is accomplished using the following properties where c is a positive number:*

- *(i) if* $|x| = c$, *then* $x = c$ *or* $x = -c$
- *(ii) if* $|x| < c$, *then* $-c < x < c$
- *(iii) if* $|x| > c$, *then* $x < -c$ *or* $x > c$

To solve $|ax + b| = d$: Replace x with $ax + b$ in property (i). Write and then solve the pair of equations,

$$ax + b = d \text{ or } ax + b = -d.$$

Example: If $|2x + 3| = 1$, then

$$2x + 3 = 1 \quad \text{or} \quad 2x + 3 = -1$$

$$2x = -2 \qquad\qquad 2x = -4$$

$$\frac{2x}{2} = \frac{-2}{2} \qquad\qquad \frac{2x}{2} = \frac{-4}{2}$$

$$x = -1 \quad \text{or} \qquad x = -2$$

KEY EXAMPLE

Solve and check: $\qquad\qquad |x + 3| = 2x$

Solution: If $\qquad\qquad |x + 3| = 2x$, then

$$x + 3 = 2x \quad \text{or} \quad x + 3 = -2x$$

$$-x + 3 = 0 \quad \text{or} \quad 3x + 3 = 0$$

$$-x = -3 \quad \text{or} \quad 3x = -3$$

$$x = 3 \quad \text{or} \quad x = -1$$

Check:

Let $x = 3$:   or   Let $x = -1$:

$$|x + 3| = 2x \quad \text{or} \quad |x + 3| = 2x$$

$$|3 + 3| = 2(3) \quad \text{or} \quad |-1 + 3| = 2(-1)$$

$$6 = 6 \quad \text{or} \quad 2 \neq -2$$

Reject this solution.

Hence, the only root of the equation is $x = 3$.

KEY EXAMPLE

Solve and check: $|2x + 1| = |2x - 9|$

Solution: In general, if $|A| = |B|$, then $A = |B|$ or $A = -|B|$ which means $A = B$ or $A = -B$. Hence, if

$$|2x + 1| = |2x - 9|,$$

then

$$2x + 1 = 2x - 9 \quad \text{or} \quad 2x + 1 = -(2x - 9)$$

$$1 = -9 \quad\quad\quad\quad 2x + 1 = -2x + 9$$

Impossible result implies
the equation has no solution.

$$2x = -2x + 8$$

$$4x = 8$$

$$x = 2$$

The check is left for you.

To solve $|ax + b| < d$:   Replace x with $ax + b$ in property (ii). Write and then solve the compound inequality

$$-d < ax + b < d.$$

If

$$|ax + b| \leq d,$$

then

$$-d \leq ax + b \leq d.$$

To solve $|ax + b| > d$:   Replace x with $ax + b$ in property (iii). Write and then solve the compound inequality,

$$ax + b < -d \text{ or } ax + b > d.$$

If

$$|ax + b| \geq d,$$

then

$$ax + b \leq -d \text{ or } ax + b \geq d.$$

KEY EXAMPLE

Solve for x:
$$\left|\frac{5x-7}{3}\right| < 1$$

Solution: If
$$\left|\frac{5x-7}{3}\right| < 1$$

then
$$-1 < \frac{5x-7}{3} < 1$$

Isolate the variable making sure to perform the same operations on both sides of the inequality:

$$(3)(-1) < 3\left(\frac{5x-7}{3}\right) < (3)(1)$$

$$-3 + 7 < 5x - 7 + 7 < 3 + 7$$

$$4 < 5x < 10$$

$$0.8 < x < 2$$

The solution is the open interval (0.8, 2).

KEY EXAMPLE

Solve for x: $\qquad\qquad 11 < |4 - 3x|$.

Solution : If $\qquad\quad 11 < |4 - 3x|$,
then $\qquad\qquad |4 - 3x| > 11$
so

$$4 - 3x < -11 \quad \text{or} \quad 4 - 3x > 11$$

$$-3x < -11 - 4 \quad \text{or} \quad 4 - 3x - 4 > 11 - 4$$

$$-3x < -15 \quad \text{or} \quad -3x < 7$$

$$x > 5 \quad \text{or} \quad x < -\frac{7}{3}$$

The solution is $x > 5$ or $x < -\dfrac{7}{3}$.

Key 13 Quadratic equations

OVERVIEW   *A **quadratic equation** in* x *is an equation that can be written in the standard form* $ax^2 + bx + c = 0 (a \neq 0)$. *Some quadratic equations can be solved by factoring the quadratic polynomial.*

Zero product rule: If, $a \cdot b = 0$ then either $a = 0$, $b = 0$, or both $a = 0$ and $b = 0$.

Solving quadratic equations by factoring: If the left side of the quadratic equation $ax^2 + bx + c = 0$ ($a \neq 0$) can be factored, then by the zero product rule each linear factor is equal to 0. The solutions to these linear equations are the roots of the quadratic equation. For example, if $x^2 + 4x = 0$, then $x(x + 4) = 0$. Since either factor can equal 0, $x = 0$ or $x + 4 = 0$. If $x + 4 = 0$, then $x = -4$. Hence, the two roots of the equation $x^2 + 4x = 0$ are $x = 0$ and $x = -4$.

KEY EXAMPLE

Solve and check: $3x^2 + 6x = 45$

Solution: Write the equation in the standard form $ax^2 + bx + c$:

$$3x^2 + 6x - 45 = 0$$

To simplify the factoring, divide through by any common numerical factors:

$$\frac{3x^2}{3} + \frac{6x}{3} - \frac{45}{3} = \frac{0}{3}$$

$$x^2 + 2x - 15 = 0$$

Factor $(x - 3)(x + 5) = 0$

Apply the zero product rule: $x - 3 = 0$ or $x + 5 = 0$

$$x = 3 \quad \text{or} \quad x = -5$$

The check is left for you.

KEY EXAMPLE

Solve for y and check: $(2y - 1)(y + 3) = 4$

Solution: Since the right side of the equation is not 0, the zero product rule cannot be used. Put the equation into standard form by multiplying the factors and then collecting all terms on the left side of the equation with 0 on the right side. The result is

$$2y^2 + 5y - 7 = 0.$$

Factoring and using the zero product rule leads to

$$2y + 7 = 0 \text{ or } y - 1 = 0.$$

Hence, the roots of the equation are $-\dfrac{7}{2}$ and 1.

The check is left for you.

Solving higher degree equations by factoring: Sometimes an equation of degree higher than 2 can be solved by factoring completely. For example, if

$$-n^3 - 5n^2 + 6n = 0,$$

then after multiplying each term by -1, n can be factored out so

$$n(n^2 + 5n - 6) = 0.$$

Factoring the quadratic polynomial gives

$$n(n + 6)(n - 1) = 0.$$

If the product of three expressions is 0, then at least one of these expressions is 0. One root is $n = 0$. If $n + 6 = 0$, then $n = -6$. If $n - 1 = 0$, then $n = 1$. The three roots of the original third degree equation are $n = 0$ or $n = -6$, or $n = 1$.

Solving equations having a quadratic form: If a nonlinear equation can be put into the form $au^2 + bu + c = 0$ where $a \neq 0$ and u is some variable expression, then the equation has a quadratic form. For example, since the equation $x^4 - 5x^2 + 4 = 0$ can be written as $(x^2)^2 - 5x^2 + 4 = 0$, it may be made to look like a familiar type of quadratic equation by substituting u for x^2. The transformed equation is $u^2 - 5u + 4 = 0$. Its roots are $u = 1$ and $u = 4$. If $u = x^2 = 1$, then $x = \pm 1$. If $u = x^2 = 4$, then $x = \pm 2$. The *four* roots of the original *fourth* degree equation are -1, $+1$, -2, and $+2$.

KEY EXAMPLE

Solve and check: $x + \sqrt{x} - 6 = 0$

Solution: Since

$$x = (\sqrt{x})^2,$$

the given equation can be written as

$$(\sqrt{x})^2 + \sqrt{x} - 6 = 0.$$

Letting $u = \sqrt{x}$, the original equation becomes

$$u^2 + u - 6 = 0$$

whose roots are $u = 2$ and $u = -3$. If

$$u = \sqrt{x} = 2,$$

then

$$x = 4.$$

If

$$u = \sqrt{x} = -3,$$

then this solution must be rejected since $\sqrt{x}$ must be nonnegative. The original equation has the single root $x = 4$. The check is left for you.

KEY EXAMPLE

Solve for y and check:

$$2\left(\frac{1}{y+1}\right)^{\frac{2}{3}} - 5\left(\frac{1}{y+1}\right)^{\frac{1}{3}} + 2 = 0$$

Solution: Make a change in variable by letting

$$u = \left(\frac{1}{y+1}\right)^{\frac{1}{3}}$$

so

$$u^2 = \left(\frac{1}{y+1}\right)^{\frac{2}{3}}$$

Thus, the original equation becomes

$$2u^2 - 5u + 2 = 0$$

whose roots are $u = \dfrac{1}{2}$ and $u = 2$. If

$$u = \left(\frac{1}{y+1}\right)^{\frac{1}{3}} = \frac{1}{2}$$

then raising both sides of this equation to the third power gives

$$\frac{1}{y+1} = \frac{1}{8}$$

so

$$y = 7$$

If

$$u = \left(\frac{1}{y+1}\right)^{\frac{1}{3}} = 2$$

then

$$\frac{1}{y+1} = 8$$

so

$$y = -\frac{7}{8}$$

The roots of the original equation are $y = 7$ and $y = -\dfrac{7}{8}$. The check is left for you.

Key 14   Quadratic inequalities

OVERVIEW   *Replacing the equal sign in a quadratic equation with an inequality symbol produces a **quadratic inequality**. The real roots of the related quadratic equation determine the endpoints of the possible solution intervals of the quadratic inequality.*

To solve a quadratic inequality by factoring: Write the quadratic inequality so that 0 is alone on the right side of the inequality. Solve the related quadratic equation. The roots of this equation, called **critical numbers**, divide the real number line into distinct **solution intervals**. If a test value from an interval satisfies the original inequality, then the interval from which it was selected belongs to the solution set.

KEY EXAMPLE

Solve for x: $\qquad\qquad x^2 - 2x < 3$

Solution: The original inequality in standard form is
$$x^2 - 2x - 3 < 0.$$
Factoring gives
$$(x + 1)(x - 3) < 0.$$
The roots of
$$(x + 1)(x - 3) = 0$$
are $x = -1$ and $x = 3$. These critical numbers divide the real number line into three intervals: $x < -1$, $-1 < x < 3$, and $x > 3$. Select a test value from each interval and determine whether it makes the inequality $(x + 1)(x - 3) < 0$ true or false. It may help to organize your work as follows:

Interval	Test Value	$(x + 1)(x - 3)$	$(x + 1)(x - 3) < 0$?
$x < -1$	$x = -2$	$(-2 + 1)(-2 - 3) = 5$	$5 < 0$ is false.
$-1 < x < 3$	$x = 0$	$(0 + 1)(0 - 3) = -3$	$-3 < 0$ is true.
$x > 3$	$x = 4$	$(4 + 1)(4 - 3) = 5$	$5 < 0$ is false.

Since the factored form of the inequality is true for a value ($x = 0$) in the interval $-1 < x < 3$, it must also be true for every number in this interval. Hence, the solution to

$$x^2 - 2x < 3$$

is

$$-1 < x < 3.$$

General form of the solution intervals: If r_1 and r_2 ($r_1 < r_2$) represent the roots of $ax^2 + bx + c = 0$ ($a \neq 0$), then the solution intervals to the related set of quadratic inequalities are given in Table 2.1. Knowing these general forms can greatly simplify matters. For example, the solution interval to $x^2 + x - 6 \leq 0$ has the general form $r_1 \leq x \leq r_2$ where r_1 and r_2 are the roots of the related quadratic equation $x^2 + x - 6 = 0$. Since the roots of this equation are $r_1 = -3$ and $r_2 = 2$, the solution to $x^2 + x - 6 \leq 0$ is $-3 \leq x \leq 2$. If the original quadratic inequality was $x^2 + x - 6 > 0$, then the solution would be $x < -3$ or $x > 2$.

Table 2.1 General Solutions to Quadratic Inequalities ($a > 0$)

Inequality	Solution Interval(s)	Graph
$ax^2 + bx + c < 0$	$r_1 < x < r_2$	
$ax^2 + bx + c \leq 0$	$r_1 \leq x \leq r_2$	
$ax^2 + bx + c > 0$	$x < r_1$ or $x > r_2$	
$ax^2 + bx + c \geq 0$	$x \leq r_1$ or $x \geq r_2$	

Key 15 Radical equations

OVERVIEW *An equation in which the variable appears under a radical sign is solved by isolating the radical and then raising each side of the equation to the power that eliminates the radical.*

Extraneous roots: Performing the same algebraic operation on both sides of an equation does not always produce an equivalent equation. Roots of the new equation that are not roots of the original equation are called **extraneous** roots. Extraneous roots may arise when both sides of an equation are multiplied or divided by the same variable expression, or when both sides of an equation are raised to the same power. For example, squaring both sides of the equation $x = 3$ produces the equation $x^2 = 9$ whose roots are -3 and 3. Since -3 is not a root of the original equation, it is an **extraneous** root.

Radical equation: In this type of equation, the variable appears underneath a radical sign.   A radical equation like

$$\sqrt{2-x} - x = 0$$

is solved by isolating the radical and then raising both sides of the equation to the power that eliminates the radical. Squaring both sides of the equation

$$\sqrt{2-x} = x$$

gives

$$2 - x = x^2$$

or, equivalently,

$$x^2 + x - 2 = 0$$

whose roots are -2 and 1. Checking both roots in the original equation shows that $x = 1$ is a root of the radical equation, but $x = -2$ is an extraneous root.

To solve $x^{p/k} = c$: Raise both sides of the equation to the power that is the reciprocal of the fractional exponent.

Example 1. If

$$x^{2/3} = -4,$$

then

$$x = (-4)^{3/2} = \sqrt{-64}$$

Since $\sqrt{-64}$ is not real, the original equation does not have a real solution.

Example 2. If

$$x^{3/5} + 9 = 1,$$

then

$$x^{3/5} = -8$$

so

$$x = (-8)^{5/3} = (\sqrt[3]{-8})^5 = (-2)^2 = -32$$

KEY EXAMPLE

Solve for x and check: $\quad \sqrt{x+4} + \sqrt{1-x} = 3$

Solution: Isolate one radical term and then eliminate this radical by raising both sides of the equation to the second power.

$$\sqrt{x+4} = 3 - \sqrt{1-x}$$

$$(\sqrt{x+4})^2 = (3 - \sqrt{1-x})^2$$

$$x+4 = 9 - 6\sqrt{1-x} + (1-x)$$

Isolate the remaining radical term and then raise both sides of the equation to the second power:

$$(2x-6)^2 = (-6\sqrt{1-x})^2$$

$$4x^2 - 24x + 36 = 36(1-x)$$

$$4x^2 + 12x = 0$$

$$4x(x+3) = 0$$

Since $4x = 0$ or $x + 3 = 0$, $x = 0$ or $x = -3$. The solution is $x = 0$ or $x = -3$. You should verify that both solutions satisfy the original radical equation.

KEY EXAMPLE

Solve for x:
$$\sqrt{(x-1)^3} - 6 = 2$$

Solution: If

$$\sqrt{(x-1)^3} - 6 = 2$$

then

$$\sqrt{(x-1)}^3 = 8.$$

Squaring each side of the equation gives

$$(x-1)^3 = 64.$$

Raise each side of the equation to the $\frac{1}{3}$ power. Thus,

$$x - 1 = 64^{\frac{1}{3}} = 4$$

Since $x - 1 = 4$, $x = 5$. The check is left for you.

Theme 3 RATIONAL EXPRESSIONS

*I*f the numerator and the denominator of a fraction are poly-nomials, then the fraction is called a **rational expression**. We always assume that whenever a variable appears in the denominator of a fraction, it cannot have a value that will make the denominator evaluate to 0. The rules for working with fractions in algebra parallel the rules for operating with fractions in arithmetic.

Key 16 Simplifying rational expressions

OVERVIEW *Rational expressions are simplified by dividing the numerator and the denominator by any factors common to both.*

Negative of a fraction: The negative of $\dfrac{a}{b}$ may be written in the following equivalent ways:

$$-\frac{a}{b}, \frac{-a}{b}, \text{ and } \frac{a}{-b}$$

Cancellation law: The cancellation law may be expressed in the following way:

$$\frac{a \cdot c}{b \cdot c} = \frac{a}{b} (b, c \neq 0)$$

To simplify a rational expression: Factor the numerator and the denominator completely. Use the cancellation law to eliminate any factor that is common to both the numerator and the denominator since their quotient is 1. Then multiply the remaining factors. Sometimes it is necessary to factor out −1 from a factor in order to cancel out like factors as in

$$\frac{2a - 2b}{bx - ax} = \frac{2(a - b)}{x(b - a)} = \frac{-2(b - a)}{x(b - a)} = -\frac{2}{x}$$

KEY EXAMPLE

Simplify

$$\frac{n - m}{m^2 - n^2}$$

Solution: Factor the denominator as the difference of two squares, $(m - n)(m + n)$. By factoring out −1 from the numerator, the quotient of a pair of like factors can be obtained.

$$\frac{n - m}{m^2 - n^2} = \frac{n - m}{(m - n)(m + n)} = \frac{-1(m - n)}{(m - n)(m + n)} = -\frac{1}{m + n}$$

KEY EXAMPLE

Write the fraction $\dfrac{x^3 - 9x}{x^2 + 4x + 3}$ in lowest terms.

Solution: Factor the numerator completely and factor the denominator as the product of two binomials.

$$\frac{x^3 - 9x}{x^2 + 4x + 3} = \frac{x(x^2 - 9)}{(x+3)(x+1)}$$

$$= \frac{x\cancel{(x+3)}(x-3)}{\cancel{(x+3)}(x+1)}$$

$$= \frac{x^2 - 3x}{x+1}$$

Key 17 Operations with rational

expressions

OVERVIEW *The rules for multiplying, dividing, and combining rational expressions are the same as those for fractions in arithmetic.*

To multiply rational expressions: If possible, factor the numerators and the denominators so that if the same factor appears in both a numerator and a denominator it can be divided out before multiplying. Write the product of the factors remaining in the numerators over the product of the factors remaining in the denominators. The result is the product of the original fractions in lowest terms.

KEY EXAMPLE

Write the product in lowest terms:

$$\frac{x^2-16}{10x} \cdot \frac{15x^3}{3x+12}$$

Solution: Factor the numerators and the denominators.

$$\frac{x^2-16}{10x} \cdot \frac{15x^3}{3x+12} = \frac{(x-4)(x+4)}{(5x)(2)} \cdot \frac{(5x)(3x^2)}{3(x+4)}$$

$$= \frac{(x-4)}{2} \cdot \frac{x^2}{1}$$

$$= \frac{x^3-4x^2}{2}$$

To divide rational expressions: Invert the second fraction and then multiply the two fractions.

KEY EXAMPLE

Write each quotient in lowest terms:

$$\text{(a)} \quad \frac{6x - 3x^2}{9x^3 - x} \div \frac{x^2 - 4}{3x^2 + 5x - 2} \qquad \text{(b)} \quad \frac{x^2 + 2xy + y^2}{xy - x} \div \frac{x^2 - y^2}{y^2 - y}$$

Solution: Change to an equivalent multiplication example by inverting the second fraction.

$$\text{(a)} \quad \frac{6x - 3x^2}{9x^3 - x} \div \frac{x^2 - 4}{3x^2 + 5x - 2} = \frac{6x - 3x^2}{9x^3 - x} \cdot \frac{3x^2 + 5x - 2}{x^2 - 4}$$

$$= \frac{3\overset{1}{\cancel{x}} \, (2 \overset{-1}{\cancel{-x}})}{\cancel{x}(\cancel{3x-1})(3x+1)} \cdot \frac{\overset{1}{\cancel{(3x-1)}} \, (\overset{1}{\cancel{x+2}})}{(\cancel{x-2})(\cancel{x+2})}$$

$$= \frac{-3}{3x + 1}$$

$$\text{(b)} \quad \frac{x^2 + 2xy + y^2}{xy - x} \div \frac{x^2 - y^2}{y^2 - y} = \frac{x^2 + 2xy + y^2}{xy - x} \cdot \frac{y^2 - y}{x^2 - y^2}$$

$$= \frac{(x + y)\overset{1}{\cancel{(x+y)}}}{x\cancel{(y-1)}} \cdot \frac{y\overset{1}{\cancel{(y-1)}}}{(\cancel{x+y})(x - y)}$$

$$= \frac{(x + y)y}{x(x - y)}$$

$$= \frac{xy + y^2}{x^2 - xy}$$

To combine rational expressions: If the fractions have like denominators, write the sum (or difference) of their numerators over their common denominator and then simplify, if possible. If the fractions have unlike denominators, change the fractions to equivalent fractions having the lowest common multiple of their denominators as their common denominator. This denominator is called the **LCD** (Least Common Denominator).

KEY EXAMPLE

Express the sum in lowest terms:

$$\frac{4x}{x^2 - 4} + \frac{x+6}{4 - x^2}$$

Solution: Obtain like denominators by multiplying the numerator and the denominator of the second fraction by -1:

$$\frac{4x}{x^2 - 4} + \frac{-(x+6)}{-(4 - x^2)} = \frac{4x}{x^2 - 4} + \frac{-(x+6)}{x^2 - 4}$$

$$= \frac{4x - (x+6)}{x^2 - 4}$$

$$= \frac{3x - 6}{x^2 - 4}$$

$$= \frac{3(x-2)}{(x-2)(x+2)}$$

$$= \frac{3}{x+2}$$

KEY EXAMPLE

Combine and express the result in lowest terms:

$$\frac{x}{x^2 + x - 6} - \frac{2}{x+3} + \frac{x-3}{x^2 - 4x + 4}$$

Solution: Factor the denominators of the first and last fractions which gives

$$x^2 + x - 6 = (x + 3)(x - 2)$$

and

$$x^2 - 4x + 4 = (x - 2)^2$$

The LCD includes all the different prime factors of the three denominators with each prime factor raised to the highest power to which it occurs in any denominator. Since the factor $(x - 2)$ is raised to the second power in the last denominator, the LCD = $(x + 3)(x - 2)^2$. Multiply each fraction by "1" in the form of the factor(s) of the LCD its denominator lacks, divided by the same factor(s).

Thus

$$\frac{x}{x^2 + x - 6} - \frac{2}{x + 3} + \frac{x - 3}{x^2 - 4x + 4}$$

$$= \frac{x}{(x + 3) \cdot (x - 2)} \cdot \frac{(x - 2)}{(x - 2)} - \frac{2}{x + 3} \cdot \frac{(x - 2)^2}{(x - 2)^2} + \frac{x - 3}{(x - 2)^2} \cdot \frac{(x + 3)}{(x + 3}$$

$$= \frac{x(x - 2)}{(x + 3)(x - 2)^2} - \frac{2(x - 2)^2}{(x + 3)(x - 2)^2} + \frac{(x - 3)(x + 3)}{(x - 2)^2(x + 3)}$$

$$= \frac{x^2 - 2x - 2(x^2 - 4x + 4) + x^2 - 9}{(x + 3)(x - 2)^2}$$

$$= \frac{6x - 17}{(x + 3)(x - 2)^2}$$

Key 18 Simplifying complex fractions

OVERVIEW   *A **complex fraction** is a fraction that contains other fractions in its numerator, denominator, or in both. Simplifying a complex fraction means eliminating any fractions from its numerator and denominator.*

To simplify a complex fraction by combining terms: Rewrite the complex fraction so a single fraction appears in its numerator and in its denominator. Then divide the numerator of the complex fraction by its denominator.

Example:

$$\frac{n - \dfrac{1}{n}}{\dfrac{1}{n} + 1} = \frac{\dfrac{n^2 - 1}{n}}{\dfrac{1 + n}{n}} = \frac{n^2 - 1}{n} \div \frac{1 + n}{n}$$

$$= \frac{n^2 - 1}{n} \times \frac{n}{1 + n}$$

$$= \frac{(n-1)(n+1)}{n} \times \frac{n}{1+n} = n - 1$$

To simplify a complex fraction by clearing fractions: Multiply each term in both the numerator and the denominator of the complex fraction by the LCD of all the fractions that the numerator and the denominator contains.

Example:

$$\frac{ab^{-1} + 1}{ab^{-1} - a^{-1}b} = \frac{\dfrac{a}{b} + 1}{\dfrac{a}{b} - \dfrac{b}{a}}$$

LCD = ab:
$$= \frac{(ab)\left(\dfrac{a}{b} + 1\right)}{(ab)\left(\dfrac{a}{b} - \dfrac{b}{a}\right)} = \frac{(ab)\left(\dfrac{a}{b} + 1\right)}{(ab)\left(\dfrac{a}{b} - \dfrac{b}{a}\right)}$$

$$= \frac{a^2 + ab}{a^2 - b^2} \qquad = \frac{a(a + b)}{(a + b)(a - b)} = \frac{a}{a - b}$$

Key 19 Fractional equations and inequalities

OVERVIEW *Equations and inequalities in which the variable appears in the denominator of a fraction are solved by clearing the fractions using the multiplication properties of equations or inequalities.*

To solve a fractional equation: Multiply each term of the equation, right and left sides, by the LCD. The solution of the transformed equation will include all the roots of the fractional equation, but may also include extraneous roots.

KEY EXAMPLE

Solve and check:

$$\frac{x}{x+2} - \frac{1}{x-2} = \frac{8}{x^2 - 4}$$

Solution: The last denominator in factored form is $(x-2)(x+2)$. Eliminate each of the denominators by multiplying each member of the equation by the LCD, $(x-2)(x+2)$. Thus,

$$\frac{x}{x+2}[(x-2)(x+2)] - \frac{1}{x-2}[(x-2)(x+2)]$$

$$= \frac{8}{(x-2)(x+2)}[(x-2)(x+2)]$$

$$x(x-2) - (x+2) = 8$$

$$x^2 - 2x - x - 2 - 8 = 0$$

$$x^2 - 3x - 10 = 0$$

$$(x-5)(x+2) = 0$$

If $x - 5 = 0$, $x = 5$. If $x + 2 = 0$, then $x = -2$. Since $x = -2$ makes the denominator of the first fraction in the original equation 0, this root is an extraneous root and must be rejected. Since $x = 5$ satisfies the fractional equation, it is a root.
The solution is $x = 5$.

To solve a fractional inequality: Multiply each term of the inequality by the square of each variable denominator. This ensures that each term is being multiplied by a positive quantity so that the direction of the inequality is not affected. For example, to solve

$$\frac{x+3}{x-4} \le 0 \text{ ,where } x \ne 4,$$

multiply each side of the inequality by $(x-4)^2$. This yields

$$(x+3)(x-4) \le 0.$$

Since the roots of the related quadratic equation are $r_1 = -3$ and $r_2 = 4$, the solution interval for the quadratic inequality is $-3 \le x \le 4$ [see Key 14]. However, for the fractional inequality x cannot be equal to 4 so its solution interval is $-3 \le x < 4$.

KEY EXAMPLE

Solve for x:

$$\frac{3x}{x+1} \ge 2$$

Solution: Multiply each side of the inequality by the positive quantity $(x+1)^2$, where $x \ne -1$, which gives

$$3x(x+1) \ge 2(x+1)^2$$

Simplify: $3x^2 + 3x \ge 2(x^2 + 2x + 1)$

Collect like terms: $x^2 - x - 2 \ge 0$

Factor: $(x-2)(x+1) \ge 0$

Since the roots of the related quadratic equation are -1 and 2, and $x \ne -1$, the solution is $x < -1$ or $x \ge 2$.

KEY EXAMPLE

Solve for x:

$$\frac{5}{x-3} \le \frac{2}{x}$$

Solution: Clear the inequality of fractions by multiplying each side by the positive quantity $x^2(x-3)^2$, where $x \ne 0, 3$. This gives

$$5x^2(x-3) \le 2x(x-3)^2 \text{ or } 5x^2(x-3) - 2x(x-3)^2 \le 0.$$

Factor out $x(x-3)$:

$$x(x-3)[\,5x-2(x-3)\,] \leq 0$$

which simplifies to

$$x(x+2)(x-3) \leq 0$$

The values of x that make the left side of the inequality 0 are 0, -2, and 3. These critical numbers divide the number line into four distinct intervals: $x \leq -2$, $-2 \leq x < 0$, $0 < x < 3$, and $x > 3$. By choosing convenient test values on each interval, and keeping in mind $x \neq 0$ and $x \neq 3$, we can verify that the intervals $x \leq -2$ and $0 < x < 3$ satisfy the inequality $x(x+2)(x-3) \leq 0$. Thus, the solution intervals for the original inequality is $x \leq -2$ or $0 < x < 3$.

Theme 4 LINEAR EQUATIONS
AND INEQUALITIES
IN TWO VARIABLES

*T*he French mathematician René Descartes (1596–1650) developed a method for representing equations in two variables graphically. This method is based on creating a **coordinate plane** by drawing a horizontal number line called the *x*-**axis** and a vertical number line called the *y*-**axis**. The point at which the two coordinate axes intersect is called the **origin**. Each point in the plane corresponds to an ordered pair of real numbers, (x, y). By locating points in the plane that satisfy a given equation, we can draw its graph. Every first degree equation has a line as its graph and, conversely, every line can be described by a first-degree equation. The line $ax + by = c$ divides the coordinate plane into two half-planes, one of which contains all points that satisfy the inequality $ax + by > c$; the other half-plane contains all those points that satisfy the inequality $ax + by < c$.

INDIVIDUAL KEYS IN THIS THEME

20	Midpoint and distance formulas
21	Slope formula
22	Equations of lines
23	Graphing linear equations in two variables
24	Graphing linear inequalities in two variables

Key 20 Midpoint and distance formulas

OVERVIEW *If the endpoints of a line segment are known, then convenient formulas can be used to determine the midpoint and length of the line segment.*

Cartesian coordinate system: A system in which ordered pairs of real numbers of the form (x, y) represent the locations of points from two perpendicular axes that intersect at the **origin**. The first member of the ordered pair (x, y), called the **x-coordinate** or **abscissa**, tells the number of units the point is located horizontally to the right $(x > 0)$ or to the left $(x < 0)$ of the origin. The second member of the ordered pair (x, y), called the **y-coordinate** or **ordinate**, gives the number of units the point is located vertically above $(y > 0)$ or below $(y < 0)$ the origin. In Figure 4.1 the point $A(2, 3)$ is graphed by starting at the origin and moving along the x-axis two units to the right, stopping, and then moving 3 units up. Similarly, point $B(-4, 5)$ is graphed by starting at the origin and moving along the x- axis four units to the *left* (since $x < 0$), and then moving up 5 units.

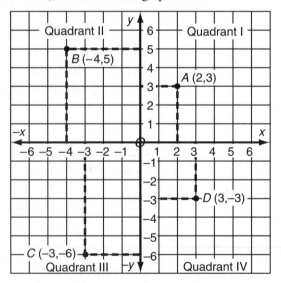

Figure 4.1 Graphing Points in the Four Quadrants

The four quadrants: The coordinate axes divide the coordinate plane into four distinct regions called **quadrants.** In Figure 4.1, the four quadrants are numbered I, II, III, and IV in counterclockwise order, beginning in the upper right corner. The signs of the *x*- and *y*-coordinates of a point determine the quadrant in which the point lies, as shown in the table that follows.

Example	Sign of Coordinates	Location of Point
$A(2, 3)$	$(+, +)$	Quadrant I
$B(-4, 5)$	$(-, +)$	Quadrant II
$C(-3, -6)$	$(-, -)$	Quadrant III
$D(3, -3)$	$(+, -)$	Quadrant IV

Midpoint Formula: If $M(\bar{x}, \bar{y})$ is the midpoint of the line segment whose endpoints are $A(x_1, y_1)$ and $B(x_2, y_2)$, as shown in Figure 4.2, then

$$M(\bar{x}, \bar{y}) = \left(\frac{x_1 + x_2}{2}, \frac{y_1 + y_2}{2} \right)$$

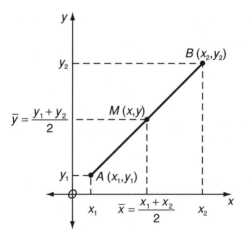

Figure 4.2 Midpoint of $\overline{AB}$

Distance formula: The distance d between points $A(x_1, y_1)$ and $B(x_2, y_2)$, as shown in Figure 4.2, is given by the equation

$$d = \sqrt{(x_2 - x_1)^2 + (y_2 - y_1)^2}$$

If points A and B lie on the same *vertical* line, then $x_2 = x_1$ and $d = |y_2 - y_1|$. If points A and B lie on the same *horizontal* line, then $y_2 = y_1$ and $d = |x_2 - x_1|$.

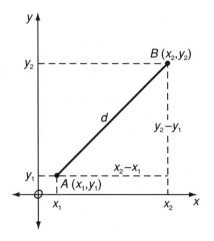

Figure 4.3 Length of $\overline{AB}$

KEY EXAMPLE

The coordinates of the endpoints of a line segment are $A(-1, 9)$ and $B(5, 1)$. Find the midpoint M and length d of $\overline{AB}$.

Solution: Let $(x_1, y_1) = (-1, 9)$ and $(x_2, y_2) = (5, 1)$.

$$M(x, y) = \left(\frac{x_1 + x_2}{2}, \frac{y_1 + y_2}{2} \right)$$

$$= \left(\frac{(-1) + 5}{2}, \frac{9 + 1}{2} \right)$$

$$= (2, 5)$$

$$d = \sqrt{(x_2 - x_1)^2 + (y_2 - y_1)^2}$$

$$= \sqrt{[5 - (-1)]^2 + (1 - 9)^2}$$

$$= \sqrt{36 + 64}$$

$$= \sqrt{100}$$

$$= 10$$

We should verify that by letting $(x_1, y_1) = (5, 1)$ and $(x_2, y_2) = (-1, 9)$, the same results are obtained.

KEY EXAMPLE

The center of a circle is located at $(7, -1)$ and the coordinates of one of the endpoints of a diameter of this circle are $(5,4)$. What are the coordinates of the other endpoint of this diameter?

Solution: The center of a circle is the midpoint of each of its diameters. Use the midpoint formula where (x, y) represents the coordinates of the unknown endpoint. Since

$$(7,-1) = \left(\frac{5+x}{2}, \frac{4+y}{2} \right)$$

$$7 = \frac{5+x}{2} \quad \text{and} \quad -1 = \frac{4+y}{2}$$

Multiplying each side of each equation by 2 gives

$$14 = 5 + x \quad \text{and} \quad -2 = 4 + y.$$

Thus, $x = 9$ and $y = -6$. The coordinates of the other endpoint of the diameter are $(9, -6)$.

Key 21 Slope formula

OVERVIEW *Moving from one point to another point on an oblique (slanted) line produces a vertical change in y-coordinates and a horizontal change in x-coordinates whose ratio provides a measure of the steepness or **slope** of the line .*

Slope formula: In general, the slope m of a nonvertical line that contains the points (x_1, y_1) and (x_2, y_2) is defined by the following equation:

$$m = \frac{\text{vertical change in } y}{\text{horizontal change in } x} = \frac{y_2 - y_1}{x_2 - x_1}$$

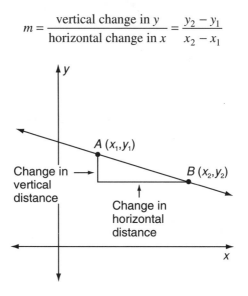

As the slope of a line increases in absolute value, the line gets "steeper."

Example: To find the slope of the line that contains the points $(1, -2)$ and $(4, 7)$, let $(x_1, y_1) = (1, -2)$ and $(x_2, y_2) = (4, 7)$. Thus,

$$m = \frac{y_2 - y_1}{x_2 - x_1} = \frac{7 - (-2)}{4 - 1} = \frac{9}{3} = 3$$

We should verify that the same slope is obtained by considering $(4, 7)$ as the "first" point and letting $(x_1, y_1) = (4, 7)$ and $(x_2, y_2) = (1, -2)$.

Positive and negative slopes: If a line rises as *x* increases, then the slope *m* of the line is positive. A line has a negative slope, if the line falls as *x* increases. These situations are shown in the accompanying figure.

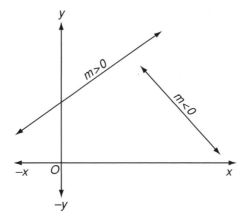

Zero slope: The slope of a horizontal line is zero.

Undefined slope: The slope of a vertical line is not defined.

Slopes of parallel lines: Parallel lines have the same slope. Conversely, different nonvertical lines that have the same slope are parallel.

Slopes of perpendicular lines: Perpendicular lines have slopes whose product is –1. Conversely, lines whose slopes have a product of –1 are perpendicular.

Proving properties of points, lines, and figures: Midpoint, distance, and slope formulas can be used to demonstrate certain properties of points, lines, and figures. Examples of some of these properties follow.

- To show that three points, *A, B,* and *C,* are collinear (lie on the same line), show that any two of the following slopes are numerically equal: slope of $\overline{AB}$, slope of $\overline{BC}$, and slope of $\overline{AC}$.
- To show that three points are the vertices of a *right* triangle:
  1. use the distance formula to find the lengths of the three sides and then show that the square of the length of the longest side is equal to the sum of the squares of the lengths of the other two sides; *or*
  2. use the slope formula to find the slopes of the three sides and then show that two of these sides have slopes whose product is –1 and, as a result, form a right angle.

- To show that four points are the vertices of a special type of quadrilateral, compute and then compare the slopes of the four sides. If exactly one pair of sides have the same slope, then these sides are parallel and the quadrilateral is a trapezoid; if two pairs of sides have the same slope, then both pairs of sides are parallel and the quadrilateral is a parallelogram. If the quadrilateral is a parallelogram and the product of the slopes of a pair of adjacent sides is -1, then the parallelogram is a rectangle. If a pair of adjacent sides of the rectangle have the same length, then the rectangle is a square.

Key 22 Equations of lines

OVERVIEW *The numerical relationship between the* x-
and y-*coordinates of each point on a line is the same and can
be expressed as a first-degree equation in* x *and* y.

Equation of a line in standard form: If a, b, and c are constants and
a and b are not both zero, then $ax + by = c$ represents the standard
form of the equation of a line.

Intercepts: An oblique line will cross both the x- and y-axis. The x-
coordinate of the point at which a line crosses the x-axis is called the
x-intercept. The y-coordinate of the point at which a line crosses the
y-axis is called the **y-intercept.**

Example: If an equation of a line is $3x - 4y = 12$, then replacing y
with 0 and solving the resulting equation for x yields $x = 4$ as the
x-intercept. Similarly, replacing x with 0 in $3x - 4y = 12$ and solving
for y produces $y = -3$ as the y-intercept.

Equation of a line in point-slope form: If m is the slope of a nonvertical line and (x_1, y_1) is a point on the line, then an equation of this
line is $y - y_1 = m(x - x_1)$. For example, if a line whose slope is 2
contains the point $(-3, 1)$, then an equation of this line is found by
letting $m = 2$ and $(x_1, y_1) = (-3, 1)$:

$$y - y_1 = m(x - x_1)$$
$$y - 1 = 2(x - (-3))$$
$$y = 2x + 7$$

Equation of a line in slope-intercept form: If m represents the slope
of a nonvertical line and b is its y-intercept, then an equation of this
line is **$y = mx + b$**. If a line has a slope of -1 and a y-intercept of 4,
then an equation of this line is $y = -x + 4$.

Example: To find the slope and y-intercept of the line

$$10x + 2y = 5,$$

put the equation in $y = mx + b$ form. This gives

$$y = -5x + \frac{5}{2}.$$

Since $m = -5$ and $b = \dfrac{5}{2}$, the slope of the line is -5 and its y-intercept is $\dfrac{5}{2}$.

Comparing equations of lines: Comparing the slopes of lines whose equations are in slope-intercept form may provide useful information about the lines.

Example 1: The lines $y = 2x - 1$ and $y = 2x + 3$ have the same slope ($m = 2$) and, as a result, are parallel.

Example 2: The lines $y = -3x$ and $y = \dfrac{1}{3}x + 7$ have slopes whose product is

$$-3\left(\frac{1}{3}\right) = -1$$

and, as a result, are perpendicular.

KEY EXAMPLE

Write an equation of a line that is parallel to the line $y + 2x = 5$ and which contains the point $(-1, 2)$.

Solution: In slope-intercept form, the given equation is $y = -2x + 5$ so its slope is -2. Since parallel lines have the same slope, the slope of the desired line must also be -2. Use the point-slope form where $m = -2$ and $(x_1, y_1) = (-1, 2)$.

$$\begin{aligned}
y - y_1 &= m(x - x_1) \\
y - 2 &= -2(x - (-1)) \\
y &= -2x
\end{aligned}$$

KEY EXAMPLE

Write an equation of the line that contains the points $(6, 0)$ and $(2, -6)$.

Solution: First find the slope of the line. Then use the point-slope form, letting (x_1, y_1) equal either of the given points.

$$m = \frac{-6 - 0}{2 - 6} = \frac{-6}{-4} = \frac{3}{2}$$

$$y - y_1 = m(x - x_1)$$

Let $(x_1, y_1) = (6, 0)$:

$$y - 0 = \frac{3}{2}(x - 6)$$

$$y = \frac{3}{2}x - 9$$

Equations of horizontal and vertical lines: A vertical line will intersect the x-axis and a horizontal line will intersect the y-axis.

- $x = a$ is an equation of a vertical line whose x-intercept is a.
- $y = b$ is an equation of a horizontal line whose y-intercept is b.

Example: An equation of a line that contains $(-3, 5)$ and is parallel to the x-axis is $y = 5$. If the line contains $(-3, 5)$ and is parallel to the y-axis, then an equation of the line is $x = -3$.

Key 23 Graphing linear equations
in two variables

OVERVIEW *Graphing a line means using several points that the line contains to correctly place the line on a coordinate plane.*

To graph a line from its equation:

1. Draw the coordinate axes on graph paper;
2. Label the origin 0, the horizontal axis x, and the vertical axis y;
3. Choose an appropriate scale along each axis;
4. Determine the coordinates of at least two points that satisfy the equation of the line;
5. Graph these points and then connect them with a straight line;
6. Label the line with its equation.

Determining points that satisfy a linear equation: Any of the following methods can be used to obtain points that satisfy a linear equation whose graph is an oblique (slanted) line:

- Intercept Method: Determine from the linear equation the point at which the line crosses the x-axis by letting $y = 0$ and solving for x, and the point at which the line crosses the y-axis by letting $x = 0$ and solving for y. This method can only be used for oblique lines that do not pass through the origin.

  Example: The graph of the equation $2x + 3y = 6$ crosses the x-axis at $(3, 0)$, and the y-axis at $(0, 2)$.

- Slope-Intercept Method: Put the equation in $y = mx + b$ form. Graph $(0, b)$, the point at which the line crosses the y-axis. Use the slope to locate at least one additional point.

  Example: The graph of the equation $y = 3x + 1$ crosses the y-axis at $(0, 1)$. Since the slope of the line is $3 \left(= \dfrac{1}{3} \right)$, you can obtain another point on this line by starting at $(0, 1)$ and moving vertically up 3 units and then moving horizontally to the right 1 unit. The coordinates of this point are $(0 + 1, 1 + 3) = (1, 4)$.

- Three-Point Method: Solve the given equation for y, if necessary. Determine the coordinates of three points on the line by replacing x with three convenient values and then determining the corresponding values of y. Although two points determine a line, the third point serves as a "check" point. This method is illustrated in the following Key Example.

KEY EXAMPLE

Draw the graph of $y - 2x = 4$.

Solution:

1. Solve the given equation for y: $\quad y = 2x + 4$
2. Find the corresponding values of y for three convenient values of x, say $x = -2$, 0, and 2.

x	$2x + 4 = y$	(x, y)
-2	$2(-2) + 4 = 0$	$(-2, 0)$
0	$2(0) + 4 = 4$	$(0, 4)$
2	$2(2) + 4 = 8$	$(2, 8)$

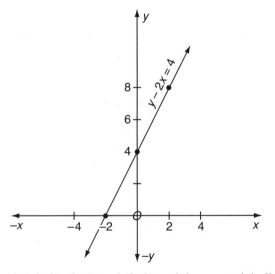

3. Graph $(-2, 0)$, $(0, 4)$, and $(2, 8)$, and draw a straight line through them. If one of the points does not lie on the line, then one of the points either has the wrong coordinates or has been graphed incorrectly.
4. Label the line with its equation.

Key 24 Graphing linear inequalities

in two variables

OVERVIEW *To graph a two-variable linear inequality, first graph the corresponding equation. Then determine which side of the line contains the set of points that satisfy the original inequality.*

To graph a linear inequality: Using any convenient method, graph the corresponding linear equation which serves as a boundary line. Pick a convenient test point that lies on one side of the line. If the coordinates of the test point satisfy the inequality, then the solution contains all points on the same side of the line. Otherwise, the solution contains all points on the opposite side of the line. If the inequality relation is < or >, then draw a broken (dashed) boundary line. If the inequality relation is ≤ or ≥, then draw a solid boundary line which indicates that points on the line are also included in the solution.

KEY EXAMPLE

Graph $2y - x < 8$.

Solution: Graph $2y - x = 8$. If $x = 0$, then $y = 4$ so the y-intercept is 4. If $y = 0$, then $x = -8$ so the x-intercept is -8. Draw a broken line since points on the line do not satisfy the inequality. (*Note* : if the original inequality was $2y - x \le 8$, then a solid line would be drawn to indicate that points on the line are in the solution.)

Pick (0, 0) as a test point:

$$2y - x < 8$$
$$?$$
$$2(0) - 0 < 8$$
$$0 < 8 \quad \text{True.}$$

Since the test point satisfies the inequality, shade in the region on the side of the line where the point lies.

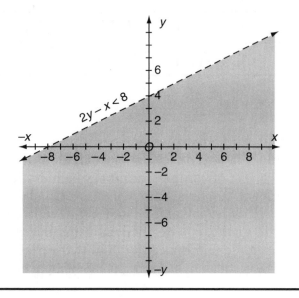

Theme 5 COMPLEX NUMBERS AND QUADRATIC EQUATIONS

A square root of a negative number is given meaning by defining a new type of number called an **imaginary number**. The set of **complex numbers** includes real numbers, imaginary numbers, and their sums. Quadratic equations that cannot be solved by factoring have irrational or imaginary roots that can be obtained using the **quadratic formula**. Every quadratic equation in two variables, only one of which is squared, has a graph called a **parabola**.

INDIVIDUAL KEYS IN THIS THEME	
25	Complex numbers
26	Solving quadratic equations by completing the square
27	Solving quadratic equations by formula
28	Sum and product of the roots
29	Graphing quadratic equations in two variables

Key 25 Complex numbers

OVERVIEW *Although $\sqrt{-1}$ satisfies the equation $x^2 = -1$, the equation has no real roots. As a result, mathematicians expanded our number system by introducing the imaginary unit i and defining it so that $i = \sqrt{-1}$ and $i^2 = -1$.*

Pure imaginary numbers: Pure imaginary numbers have the form bi where b is a nonzero real number and $i = \sqrt{-1}$. Square roots of negative numbers other than -1 can be expressed as pure imaginary numbers by factoring out $\sqrt{-1}$ and replacing it with i.

Example 1. $\sqrt{-9} = \sqrt{9}\sqrt{-1} = 3i$

Example 2. $\sqrt{-50} = \sqrt{50}\sqrt{-1} = \sqrt{25}\sqrt{2}i = 5\sqrt{2}\,i$

To combine two imaginary numbers: Combine the numerical factors of i, as in $2i + 5i = 7i$.

To multiply two imaginary numbers: Use the rule,

$$ci \cdot di = c\,d\,(i^2) = -cd$$

Example:

$$2i \cdot 5i = 10i^2 = -10$$

To multiply two imaginary numbers that are in radical form, express each number in terms of i and then multiply.

Example: $\sqrt{-4} \cdot \sqrt{-25} = (2i)(5i) = 10i^2 = -10$

Notice that $\sqrt{-4} \cdot \sqrt{-25} \neq \sqrt{100}$. The rule $\sqrt{a} \cdot \sqrt{b} = \sqrt{ab}$ holds when at least one of a and b is nonnegative 0, but is not valid if a and b are both negative.

Simplifying powers of i: Any nonnegative integer power of i can be reduced to ± 1 or $\pm i$. We know that $i^0 = 1$, $i^1 = i$, $i^2 = -1$, and $i^3 = i^2 i = -i$. Higher integer powers of i follow the same cyclic pattern that repeats every four integers. Thus, to evaluate i^n where n is a positive integer greater than 3, divide n by 4. Using the remainder as the new power of i, simplify. For example, $i^{14} = i^3$ since dividing 14 by 4 gives a remainder of 3. Thus, $i^{14} = i^3 = -1$.

Complex number: A complex number is the sum of a real number and a pure imaginary number. The complex number $4 + 3i$ is in **standard form** since it has the form $a + bi$ where a and b are real numbers and $i = \sqrt{-1}$. Every real number is a complex number. For example, since 4 can be written as $4 + 0i$, 4 is a complex number. Hence, the set of real numbers is a subset of the set of complex numbers.

Operations with complex numbers: Arithmetic operations with complex numbers are defined so that the field properties of real numbers also hold for the set of complex numbers. In performing arithmetic operations, complex numbers are treated like binomials, as in:

$$(2 + 3i) + (4 - 5i) = (2 + 4) + (3i - 5i) = 6 - 2i$$

Two complex numbers can be multiplied using FOIL.
Example:

$$
\begin{array}{cccc}
\mathbf{F} & \mathbf{O} & \mathbf{I} & \mathbf{L} \\
\end{array}
$$

$$(2 + 3i)(4 - 5i) = 2 \cdot 4 + (2)(-5i) + (3i)(4) + (3i)(-5i)$$

$$= 8 - 2i - 15i^2$$

$$= 8 - 2i - 15(-1)$$

$$= 23 - 2i$$

Complex conjugates: Complex conjugates are two complex numbers such as $a + bi$ and $a - bi$ that are the sum and difference of the same two terms. The product of a pair of complex conjugates is always a real number since

$$(a + bi)(a - bi) = (a^2) + (-abi) + (abi) + (-b^2)(i^2)$$

$$= a^2 + 0 + (-b^2)(-1)$$

$$= a^2 + b^2$$

Dividing complex numbers: To express the quotient of $\dfrac{c + di}{a + bi}$ as a complex number in standard form, multiply the fraction by "1" in the form of the complex conjugate of the denominator divided by itself. If $a = 0$, then multiply the numerator and the denominator by i.

KEY EXAMPLE

Express the quotients in standard form:

(a) $\dfrac{2 - 7i}{2i}$ (b) $\dfrac{5i}{1 + 4i}$

Solution:

(a) $\dfrac{2 - 7i}{2i} = \dfrac{2 - 7i}{2i} \cdot \dfrac{i}{i} = \dfrac{i(2 - 7i)}{2i^2}$

Multiply each term inside the parentheses by i:

$$= \frac{2i - 7i^2}{2i^2}$$

Replace i^2 by -1:

$$= \frac{7 + 2i}{-2}$$

Divide each term of the numerator by -2:

$$= -\frac{7}{2} - i$$

(b) Multiply the numerator and the denominator by $1 - 4i$ which is the complex conjugate of $1 + 4i$:

$$\frac{5i}{1 + 4i} = \frac{5i}{1 + 4i} \cdot \frac{1 - 4i}{1 - 4i}$$

$$= \frac{5i(1 - 4i)}{1^2 + 4^2}$$

Multiply each term inside the parentheses by $5i$:

$$= \frac{5i - 20i^2}{17}$$

$$= \frac{20}{17} + \frac{5i}{17}$$

Key 26 Solving quadratic equations
by completing the square

OVERVIEW *If the quadratic polynomial of a quadratic equation in standard form cannot be factored, then the quadratic equation can be solved by writing it in the form* $(Ax + B)^2 = k$. *The solutions for* x *are obtained by taking the square root of each side of the equation.*

Solving quadratic equations by completing the square: Quadratic equations like $4x^2 - 24x + 7 = 0$ have quadratic polynomials that cannot be factored with integers. We can solve $4x^2 - 24x + 7 = 0$ by using a procedure called **completing the square**, as illustrated below.

$$4x^2 - 24x + 7 = 0$$

Step 1. Divide each term of the equation by 4, the coefficient of x^2.

$$x^2 - 6x + \frac{7}{4} = 0$$

Step 2. Subtract the constant term from both sides of the equation.

$$x^2 - 6 = -\frac{7}{4}$$

Step 3. Take one-half the coefficient of x and square it.

$$\left[\frac{1}{2}(-6)\right]^2 = 9$$

Step 4. Add the result obtained in Step 3 to both sides of the equation.

$$x^2 - 6x + 9 = -\frac{7}{4} + 9$$

Step 5. Factor the left side of the equation as the square of a binomial and simplify the right side.

$$(x - 3)^2 = \frac{29}{4}$$

Step 6. Take the square root of both sides of the equation and solve for x.

$$x - 3 = \pm\frac{\sqrt{29}}{2}$$

$$x = 3 \pm \frac{\sqrt{29}}{2}$$

Step 7. Write the two roots. $x_1 = 3 + \frac{\sqrt{29}}{2}$ and $x_2 = 3 - \frac{\sqrt{29}}{2}$

Key 27   Solving quadratic equations
by formula

OVERVIEW   *Solving the quadratic equation* $ax^2 + bx + c = 0$ $(a \neq 0)$ *using the method of completing the square gives a general formula that can be used to solve any quadratic equation.*

The quadratic formula: Using this formula, each of the roots of the quadratic equation $ax^2 + bx + c = 0$ $(a \neq 0)$ are expressed in terms of the constants a, b, and c:

$$x = \frac{-b \pm \sqrt{b^2 - 4ac}}{2a} \ (a \neq 0)$$

The quantity underneath the radical sign is called the **discriminant** and may be denoted by D. Thus, $D = b^2 - 4ac$. Since the quadratic formula can be used to find the roots of *any* quadratic equation in standard form, it should be memorized.

KEY EXAMPLE

Solve for x: $2x^2 + 3x = 1$.

Solution: Put the equation into the standard form

$$2x^2 + 3x - 1 = 0.$$

Let $a = 2$, $b = 3$ and $c = -1$. Then $D = b^2 - 4ac = 3^2 - 4(2)(-) = 17$. Next, substitute into the quadratic formula:

$$x = \frac{-b \pm \sqrt{D}}{2a} = \frac{-(3) \pm \sqrt{17}}{2(2)} = \frac{-3 \pm \sqrt{17}}{4}$$

The two roots are

$$x_1 = \frac{-3 + \sqrt{17}}{4} \quad \text{and} \quad x_2 = \frac{-3 - \sqrt{17}}{4}$$

KEY EXAMPLE

Solve for x: $x^2 + 5 = 2x$.

Solution: Put the equation into the standard form

$$x^2 - 2x + 5 = 0.$$

Let $a = 1$, $b = -2$ and $c = 5$. Then $D = b^2 - 4ac = -16$. Substitute into the quadratic formula

$$x = \frac{-b \pm \sqrt{D}}{2a} = \frac{-(2) \pm \sqrt{-16}}{2(1)} = \frac{2}{2} \pm \frac{4i}{2} = 1 \pm 2i$$

The two roots are $x_1 = 1 + 2i$ and $x_2 = 1 - 2i$

Describing the type of roots: Without solving a quadratic equation having the form $ax^2 + bx + c = 0$, the type of roots it will have can be predicted by examining the discriminant $D = b^2 - 4ac$:

- If $D < 0$, the roots are imaginary and occur as a pair of complex conjugates.
- If $D = 0$, the roots are real, rational, and equal.
- If $D > 0$, the roots are real and unequal. If $b^2 - 4ac$ is a perfect square, the roots are rational; otherwise the roots are irrational and occur in conjugate pairs.

KEY EXAMPLE

Without solving $3x^2 - 11x = 4$, describe the nature of its roots.

Solution: The quadratic equation in standard form is $3x^2 - 11x - 4 = 0$. Let $a = 3$, $b = -11$, and $c = -4$. Then $D = b^2 - 4ac = 169$. Since 169 is positive and a perfect square, the roots are real, rational, and unequal.

KEY EXAMPLE

Find the smallest integer value of k such that the roots of the quadratic equation $x^2 - 5x + k = 0$ are complex.

Solution: For the equation $x^2 - 5x + k = 0$, let $a = 1$, $b = -5$, and $c = k$. Then $D = b^2 - 4ac = 25 - 4k$. If the roots are complex, then the discriminant is less than 0. Hence, $25 - 4k < 0$. Solving for k yields the inequality $k > 6\frac{1}{4}$. The *smallest integer* value of k that satisfies the inequality is 7. Thus, $k = 7$.

Key 28   Sum and product of the roots

OVERVIEW   *From the quadratic formula we know that one root is*

$$\frac{-b + \sqrt{b^2 - 4ac}}{2a}$$

and the other root is

$$\frac{-b - \sqrt{b^2 - 4ac}}{2a}$$

Adding and multiplying these expressions produce convenient formulas that show the relationships between the coefficients of a quadratic equation in standard form and the sum and product of its roots.

Sum and product of the roots: If x_1 and x_2 represent the two roots of the quadratic equation $ax^2 + bx + c = 0$ ($a \neq 0$), then:

$$\text{Sum of Roots} = x_1 + x_2 = -\frac{b}{a}$$

and

$$\text{Product of Roots} = x_1 \cdot x_2 = \frac{c}{a}$$

Furthermore, dividing each term of $ax^2 + bx + c = 0$ by the coefficient of x^2 produces the equation

$$x^2 + \frac{b}{a}x + \frac{c}{a} = 0$$

which has the general form

$$x^2 + [-(\textit{sum of roots})]\, x + (\textit{product of roots}) = 0$$

KEY EXAMPLE

If a root of the equation $x^2 + px + q$ is $2 + \sqrt{5}$, find the values of p and q.

Solution: Since irrational roots of quadratic equations occur in conjugate pairs, the other root is $2 - \sqrt{5}$. Hence, the sum of the two roots is 4 and their product is $(2 + \sqrt{5})(2 - \sqrt{5}) = 4 - 5 = -1$. Since p, the coefficient of x in the given quadratic equation, is the negative of the sum of the roots and q equals the product of the roots, $p = -4$ and $q = -1$.

KEY EXAMPLE

Write a quadratic equation whose roots are $3 + i$ and $3 - i$.

Solution: A quadratic equation whose roots are known can be formed by writing $x^2 + [-(\text{sum of roots})] x + (\text{product of roots}) = 0$. Since the sum of the given roots is 6 and their product is $(3 + i)(3 - i) = 9 - i^2 = 10$, the quadratic equation that has $3 + i$ and $3 - i$ as its roots is $x^2 - 6x + 10 = 0$.

KEY EXAMPLE

One root of the equation $2x^2 + kx - 3 = 0$ is $\dfrac{1}{2}$. Find the other root and the value of k.

Solution: In the equation $2x^2 + kx - 3 = 0$, $a = 2$, $b = k$, and $c = -3$. The product of the roots is $\dfrac{c}{a} = \dfrac{-3}{2}$ and the given root is $\dfrac{1}{2}$. The other root must be -3 since the product of -3 and $\dfrac{1}{2}$ is $\dfrac{-3}{2}$. Substitute -3 for x in the original equation and then solve for k.

$$2(-3)^2 + k(-3) - 3 = 0$$

$$18 - 3k - 3 = 0$$

$$-3k = -15$$

$$k = \frac{-15}{-3} = 5$$

Thus, the second root is -3 and the value of k is 5.

Key 29 Graphing quadratic equations
in two variables

OVERVIEW *The graph of the quadratic equation* $y = ax^2 + bx + c$ *($a \neq 0$) is a curve called a **parabola** that has a vertical axis of symmetry. The point at which the axis of symmetry intersects the parabola is called the **vertex** or **turning point**.*

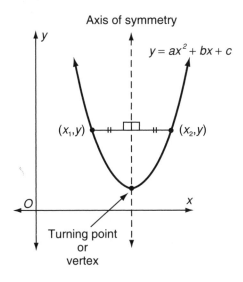

Axis of symmetry

$y = ax^2 + bx + c$

(x_1, y) (x_2, y)

O x

Turning point
or
vertex

Vertical parabola: The graph of $y = ax^2 + bx + c$ ($a \neq 0$) is a parabola with a vertical axis of symmetry. An equation of this axis of symmetry is $x = \dfrac{-b}{a}$. Thus, the x-coordinate of the turning point is $\dfrac{-b}{2a}$.

Substituting this value for x in $y = ax^2 + bx + c$ gives the y-coordinate of the turning point. (See Figure 5.1.) The sign of a, the coefficient of x^2, determines whether the parabola opens "up" or "down." Observe in Figure 5.1 the following:

- If $a > 0$, the parabola opens up (can "hold water") and the turning point is a *minimum* point on the parabola.
- If $a < 0$, the parabola opens down ("spills water") and the turning point is a *maximum* point on the parabola.

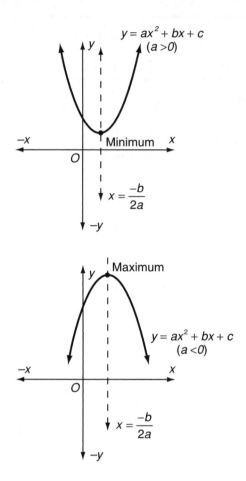

Figure 5.1 The Effect of the Sign of the Coefficient of x^2 on the Parabola.

Graph of $y = ax^2$ ($a \neq 0$): The parabola $y = ax^2$ has the origin as its turning point and the y-axis as its axis of symmetry.

- As $|a|$ increases, the parabola becomes narrower horizontally. Thus, the graph of $y = 2x^2$ is narrower than the graph of $y = x^2$.
- As $|a|$ decreases, the parabola becomes broader horizontally.

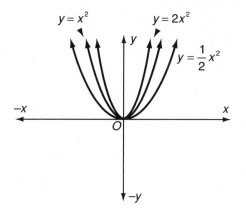

Graph of $y = ax^2 + c$ ($a \neq 0$): The parabola $y = ax^2 + c$ is the graph of $y = ax^2$ shifted vertically $|c|$ units.

1. If $c > 0$, the shift is up.
2. If $c < 0$, the shift is down.

The accompanying figure illustrates the effect of c for the cases in which $c = 2$ and $c = -2$.

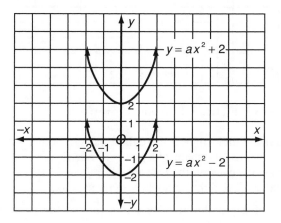

To sketch $y = ax^2 + bx + c$**:** Determine and graph the axis of symmetry and the turning point. Then graph several pairs of corresponding points that are on either side of the axis of symmetry and that are the same distance from it. For example, the parabola whose equation is $y = x^2 - 4x + 1$,

- The equation of the axis of symmetry is $x = \dfrac{-b}{2a} = \dfrac{-(-4)}{2(1)} = 2$. The y-coordinate of the vertex is found by replacing x with 2 in $y = x^2 - 4x + 1$, thus obtaining $y = -3$. Graph the turning point $(2, -3)$ and the vertical line of symmetry, $x = 2$, that passes through it.

- Choose corresponding pairs of points whose x-coordinates are 1, 2, and 3 units on either side of the x-coordinate of the turning point. Substituting these values of x in the equation of the parabola gives three pairs of symmetric points: $(1, -2)$ and $(3, -2)$; $(0, 1)$ and $(4, 1)$; $(-1, 6)$ and $(5, 6)$.

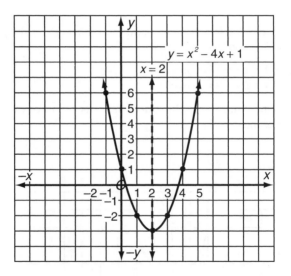

Figure 5.2 Graph of $y = x^2 - 4x + 1$

- Graph and then connect these three pairs of points with a smooth curve that has the shape of a parabola. See Figure 5.2.
- Use arrowheads on the parabola to indicate that it rises without bound, and label the parabola with its equation.

Intercepts of a parabola: The y-intercept of the parabola whose equation is $y = ax^2 + bx + c$ is c. The x-intercepts, if any, are the real roots of the quadratic equation $ax^2 + bx + c = 0$. If the roots of

$ax^2 + bx + c = 0$ are equal, then the parabola will be tangent to the x-axis. That is, the parabola will intersect the x-axis in exactly one point. If the roots of $ax^2 + bx + c = 0$ are complex, then the parabola has no x-intercepts and, as a result, will not intersect the x-axis.

Horizontal parabola: The graph of $x = ay^2 + by + c$ $(a \neq 0)$ is a parabola with a horizontal axis of symmetry. An equation of this axis of symmetry is $y = \dfrac{-b}{2a}$. As illustrated in Figure 5.3, if $a > 0$, then the parabola opens to the right; if $a < 0$, the parabola opens to the left.

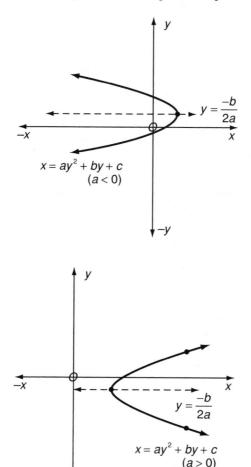

Figure 5.3 Parabolas with a horizontal axis of symmetry

Theme 6 FUNCTIONS AND THEIR GRAPHS

*F*unctions are powerful mathematical tools for describing how two variables are related to each other. A function may take many forms. Regardless of the form it takes, a function must satisfy the following two conditions:

- there is a "rule" that tells how each of the possible values of one variable is matched with a value of the other variable. This rule may be given in words, in a diagram, as a list of ordered pairs, as a graph, or as an equation.
- no two of its ordered pairs of the form (*x, y*) have the same *x*-value but different *y*-values.

Key 30 Functions

OVERVIEW *A function may take the form of a set of ordered pairs, a graph, or an equation. Regardless of the form it takes, a function must obey the condition that no two of its ordered pairs have the same first member with different second members.*

Relation: A set of ordered pairs of the form (x, y) is called a relation.

Function: A relation in which no two ordered pairs have the same x-value but different y-values is called a function. Functions are usually designated by lower-case letters such as f, g, and h. For example, if $f = \{(-3, 9), (0, 0), (3, 9)\}$ and $g = \{(4, -2), (4, 2)\}$ then f is a function and g is *not* a function.

Domain and Range: The **domain** of a function is the set of all the first members of its ordered pairs, and the **range** of a function is the set of all second members of its ordered pairs. If function $f = \{(a, A), (b, B), (c, C)\}$, then its *domain* is $\{a, b, c\}$ and its *range* is $\{A, B, C\}$.

Functions as mappings: A function may be viewed as a *mapping* or a pairing of the elements of one set with elements of a second set such that each element of the first set, the domain, is paired with exactly one element of the second set, called the **codomain**. For example, if a function f maps $\{a, b, c\}$ into $\{A, B, C, D\}$ such that $a \rightarrow A$ (read "a is mapped into A"), $b \rightarrow B$, and $c \rightarrow C$, then the domain is $\{a, b, c\}$ and the codomain is $\{A, B, C, D\}$. Since a is paired with A in the codomain, A is called the *image* of a. Each element of the codomain that corresponds to an element of the domain is called the **image** of that element. The set of image points, $\{A, B, C\}$, is called the **range**. Thus, the range is a subset of the codomain.

Onto mappings: Set A is mapped *onto* set B if each element of set B is the image of an element of set A. Thus, every function maps its domain *onto* its range.

Describing a function by an equation: The rule by which each x-value gets paired with the corresponding y-value may be specified by an equation. For example, the function described by the equation $y = x + 1$ requires that for any choice of x in the domain, the corresponding range value is $x + 1$. Thus, $2 \rightarrow 3$, $3 \rightarrow 4$, and $4 \rightarrow 5$.

Restricting domains of functions: Unless otherwise indicated, the domain of a function is assumed to be the largest possible set of real numbers. Thus, the domain of $y = \dfrac{x}{x^2 - 4}$ is the set of all real numbers *except* ± 2 since for these values of x the denominator is 0. The domain of $y = \sqrt{x-1}$ is the set of real numbers greater than or equal to 1 since for any value of x less than 1, the square root radical has a negative radicand so the radical does not represent a real number.

KEY EXAMPLE

Determine which of the following relations, if any, describe functions:

(a) $y = \sqrt{x}$ (b) $y = x^3$ (c) $y > x$ (d) $x = y^2$

Solution: The equations in (a) and (b) produce exactly one value of y for each value of x. Hence, equations (a) and (b) describe functions. In (c), $y > x$ does *not* represent a function since it contains ordered pairs such as (1, 2) and (1, 3) in which the same value of x is paired with different values of y. In (d), the equation $x = y^2$ is not a function since ordered pairs such as (4, 2) and (4, −2) satisfy the equation but have the same value of x paired with different values of y.

Function notation: For any function f, the value of y that corresponds to a given value of x is denoted by $f(x)$. If $y = 5x - 1$, then $f(2)$, read as "f of 2," represents the value of y when $x = 2$. Thus,

$$f(2) = 5 \cdot 2 - 1 = 9.$$

If $x = 3$, then

$$f(3) = 5 \cdot 3 - 1 = 14.$$

In an equation that describes function f, $f(x)$ may be used in place of y, as in $f(x) = 5x - 1$. If $y = f(x)$, then y is said to be a function of x. Since the value of y depends on the value of x, y is called the **dependent variable** and x is called the **independent variable**.

Using variable names other than x and y: Functions may be expressed in terms of variables other than x and y. For example, the area, A, of a circle depends on the radius, r, of the circle. If the name of this function is g, then $A = g(r)$. For function g, the independent variable is r and the dependent variable is A. If a function h is described by the equation $p = -q^2 + 23q + 47$, then p is a function of q. Hence, $p = h(q)$, where q is the independent variable and p is the dependent variable.

KEY EXAMPLES

1. If $f(x) = (x + 1)^{3/2}$, find $f(8)$.

Solution: $\qquad\qquad f(8) = (8+1)^{3/2} = (\sqrt{9})^3 = 27$

2. If $g(t) = t^2$, find $g(x + 1) - g(x)$.

Solution: Since

$$g(x) = (x + 1)^2 = x^2 + 2x + 1 \text{ and } g(x) = x^2,$$

$$g(x + 1) - g(x) = (x^2 + 2x + 1) - (x^2) = 2x + 1.$$

3. If $h = \{(-2, 5), (-1, 2), (0, 1), (1, 2), (2, 5)\}$, find $h(1)$.

Solution: The function value $h(1)$ represents the y-value in h the ordered pair whose x-value is 1. Since h contains $(1, 2)$, $h(1) = 2$.

Forming new functions using arithmetic operations: Two functions may be added, subtracted, multiplied, or divided to produce a third function whose domain is the set of values that are common to the domains of the original two functions.

Example: If $f(x) = x^2$ and $g(x) = x - 1$, then

$$f(x) + g(x) = x^2 + x - 1,$$

$$f(x) - g(x) = x^2 - x + 1,$$

$$f(x) \cdot g(x) = x^3 - x^2,$$

and

$$\frac{f(x)}{g(x)} = \frac{x^2}{x - 1},$$

provided $x \neq 1$.

Key 31 Vertical and horizontal line tests

OVERVIEW *There are simple tests for telling whether a graph is a function and if it is a function, whether the same value of* y *is used in different order pairs.*

Vertical line test: A graph may or may not represent a function. A graph represents a function only if there is no vertical line that can be drawn that intersects the graph in more than one point. The graph in Figure 6.1 does not represent a function, while the graph in Figure 6.2 describes a function.

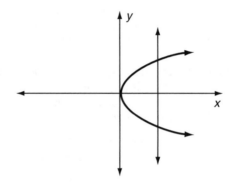

Figure 6.1 Graph is not a function.

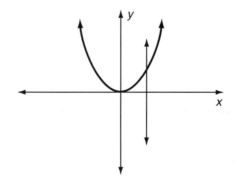

Figure 6.2 Graph is a function.

One-to-one function: A function is one-to-one if different values of x are never paired with the same value of y. If $f = \{(-3, 9), (3, 9)\}$ and $g = \{(-2, -8), (2, 8)\}$, then f is not a one-to-one function, while g is a one-to-one function. The ordered pairs for function f lie on the same horizontal line.

Horizontal line test: A function is a one-to-one function if no horizontal line intersects its graph in more than one point. The graph in Figure 6.2 fails the horizontal line test. Hence, $f(x) = x^2$ is not a one-to-one function. The accompanying graph passes both the vertical and horizontal line tests and is, therefore, a one-to-one function.

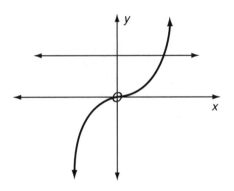

Figure 6.3 Graph is a 1:1 function.

Key 32 Symmetry and functions

OVERVIEW *A graph is* symmetric *with respect to a line if the line divides the figure into two parts such that when the coordinate plane is "folded" along the line, the two parts of the graph coincide. Some graphs have one of the coordinate axes as a line of symmetry.*

Definition and test of symmetry with respect to a coordinate axis:
Figures 6.4 and 6.5 show graphs with y-axis and x-axis symmetry respectively. The graph of an equation has y-axis or x-axis symmetry if the following conditions are satisfied:

- **y-axis symmetry** exists if whenever (x, y) is on the graph, $(-x, y)$ is also on the graph. To test for y-axis symmetry without graphing an equation, replace x with $-x$. If this produces an equivalent equation, then the graph of this equation is symmetric with respect to the y-axis.

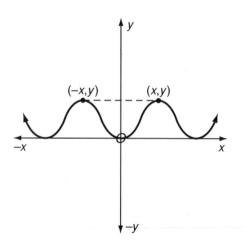

Figure 6.4 *y-axis symmetry*

Example: Replacing x with $-x$ in the equation $y = x^2 + 1$ gives $y = (-x)^2 + 1 = x^2 + 1$. Since this substitution does not change the original equation, the graph of $y = x^2 + 1$ is symmetric with respect to the y-axis.

- **x-axis symmetry** exists if whenever (x, y) is on the graph, $(x, -y)$ is also on the graph. To test for x-axis symmetry without graphing an equation, replace y with $-y$. If this produces an equivalent equation, then the graph of this equation is symmetric with respect to the x-axis.

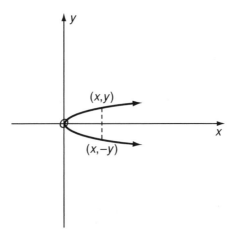

Figure 6.5 x-axis symmetry

Example 1: Replacing y with $-y$ in the equation $x = y^2 - 3$ gives $x = (-y)^2 - 3 = y^2 - 3$. Since this substitution does not change the original equation, the graph of $x = y^2 - 3$ is symmetric with respect to the x-axis.

Example 2: The circle whose equation is $x^2 + y^2 = 16$ has both x-axis and y-axis symmetry since $(-x)^2 + (-y)^2 = x^2 + y^2 = 16$.

Definition and test for origin symmetry: A graph has origin symmetry if whenever (x, y) is on the graph, $(-x, -y)$ is also on the graph. The graph in Figure 6.6 has origin symmetry. To test for origin symmetry without graphing an equation, replace x with $-x$ and y with $-y$. If this produces an equivalent equation, then the graph of this equation is symmetric with respect to the origin.

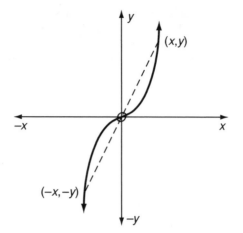

Figure 6.6 Origin symmetry

Example: The graph of $y = \dfrac{8}{x}$ ($x \neq 0$) has origin symmetry since

replacing x and y with their opposites gives $-y = \dfrac{8}{-x}$ which, after

multiplying both sides of the equation by -1, can be written as

$y = \dfrac{8}{x}$.

Even function: An even function has the property that $f(-x) = f(x)$ for all x in the domain of f. Thus, the graph of an even function has y-axis symmetry.

Example: The function $f(x) = x^4 - x^2 + 3$ is an even function since

$$f(-x) = (-x)^4 - (-x)^2 + 3 = x^4 - x^2 + 3.$$

Odd function: An odd function has the property that $f(-x) = -f(x)$ for all x in the domain of f. Thus, the graph of an odd function has origin symmetry.

Example 1: The function $f(x) = x^3 - x$ is an odd function since

$$f(-x) = (-x)^3 - (-x) = -x^3 + x = -(x^3 - x) = -f(x).$$

Example 2: The functions $f(x) = x^2 + 2x$ and $f(x) = x^3 + 1$ are neither odd nor even.

Key 33 Graphs of special functions

OVERVIEW *Some types of functions occur so frequently in mathematics that we should know their graphs.*

Identity function: The identity function is the linear function $f(x) = x$, the graph of which contains the origin and is equidistant from the coordinate axes, as shown in Figure 6.7.

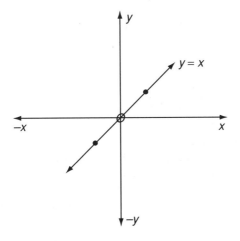

Figure 6.7   Graph of $y = x$

Reciprocal function: The graph of the reciprocal function, $f(x) = \dfrac{1}{x}$ (or $xy = 1$) is called an **equilateral (rectangular) hyperbola** (see Figure 6.8). An equilateral hyperbola consists of two branches that are located in quadrants I and III. The domain and range consist of all real numbers except x, $y \neq 0$. The coordinate axes are **asymptotes** of the curve since each branch gets closer and closer to a coordinate axis, but never touches it. The function is *odd* so its graph is symmetric with respect to the origin. In general, the graph of $f(x) = \dfrac{k}{x}$ (or $xy = k$), where k is some nonzero constant, is an equilateral hyperbola. The sign of k determines the pair of opposite quadrants in which the two branches are located.   If $k > 0$, then the branches are in quadrants I and III; if $k < 0$, the branches are in quadrants II and IV.

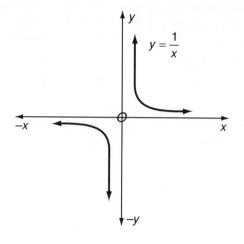

Figure 6.8 Graph of $y = \dfrac{1}{x}$

Absolute value function: The domain of the absolute value function $f(x) = |x|$ is the set of real numbers and its range is the set of non-negative real numbers. The graph, as shown in Figure 6.9, consists of the rays $y = x$ and $y = -x$ whose common endpoint is the origin. The function is *even* so its graph is symmetric with respect to the y-axis.

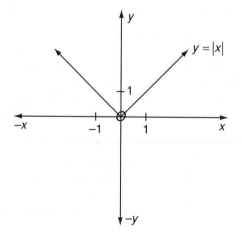

Figure 6.9 Graph of $y = |x|$

Square root function: The domain and range of the square root function $f(x) = \sqrt{x}$ are limited to the set of real nonnegative numbers. In Figure 6.10, the graph starts at the origin and gradually rises in the first quadrant.

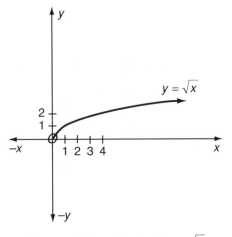

Figure 6.10 Graph of $y = \sqrt{x}$

Cube function: The domain and range of the cube function $f(x) = x^3$ is the set of real numbers. The graph, as shown in Figure 6.11, is steeper than the graph of $f(x) = x^2$ and is restricted to the first and third quadrants. Since it is an *odd* function, the graph is symmetric with respect to the origin.

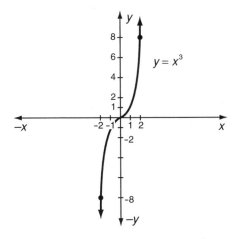

Figure 6.11 Graph of $y = x^3$

Greatest integer [step] function: This function is denoted by **[x]**, read as "the greatest integer less than or equal to x," and is defined by the equation

$$f(x) = [x] = n$$

where n is an integer such that $n \leq x < n + 1$. For example, $[3.9] = 3$ since $3 \leq 3.9 < 4$. That is, 3 is the greatest integer ≤ 3.9. To illustrate further, $[0.8] = 0$ because $0 \leq 0.8 < 1$. Also, $[-1.7] = -2$ since $-2 \leq -1.7 < -1$. Figure 6.12 shows the graph of $y = [x]$ on $-2 \leq x < 4$. Since the range of this function is restricted to integer values while the domain consists of real numbers, the graph shown in Figure 6.12 is discontinuous and consists of a rising set of horizontal "steps."

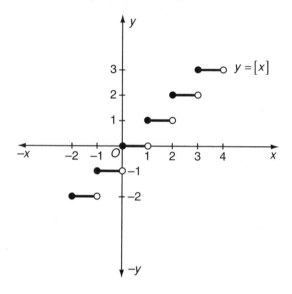

Figure 6.12   Graph of $y = [x]$

Piecewise defined functions: Functions whose definitions change over different intervals on their domains. The graph of

$$f(x) = \begin{cases} x^2, & \text{if } x < 0 \\ \sqrt{x}, & \text{if } x \geq 0 \end{cases}$$

is shown. If $x < 0$, then $f(x)$ is defined as x^2, and for $x \leq 0$, $f(x)$ is defined as $\sqrt{x}$. Thus, $f(-4) = (-4)^2 = 16$ and $f(4) = \sqrt{4} = 2$.

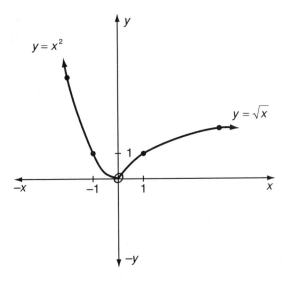

Example: The graph of

$$g(x) = \begin{cases} -x^2 + 1, & x \le 1 \\ 3x, & x > 1 \end{cases}$$

is shown. The graph is not continuous since it has a "jump" discontinuity at the "crossover point" at $x = 1$.

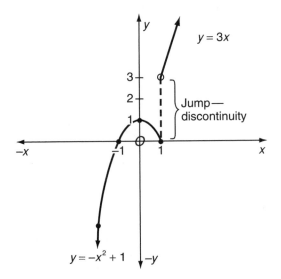

Key 34 Shifts of graphs of functions

OVERVIEW   *Two functions may be related so that the graph of one function can be obtained by shifting the graph of a related but simpler function.*

Vertical shift: If k is a constant and $g(x) = f(x) + k$, then the graph of $g(x)$ is the graph of $f(x)$ shifted *vertically* $|k|$ units. The graph of $g(x)$ is shifted *up* if $k > 0$, and is shifted *down* if $k < 0$. For example, the graph of $g(x) = x^2 + 3$ can be obtained by shifting the graph of $f(x) = x^2$ vertically up 3 units so its turning point is shifted from $(0, 0)$ to $(0, 0 + 3) = (0, 3)$.   The graph of $g(x) = x^2 - 1$, is the parabola $y = x^2$ shifted vertically down 1 unit so its turning point is $(0, -1)$.

Horizontal shift: If h is a constant and $g(x) = f(x - h)$, then the graph of $g(x)$ is the graph of $f(x)$ shifted horizontally $|h|$ units. The graph of $f(x)$ is shifted to the right if $h > 0$, and is shifted to the left if $h < 0$.

KEY EXAMPLE

Sketch the graphs of: (a) $f(x) = \sqrt{x - 1}$ (b) $f(x) = \sqrt{x} - 1$

Solution: (a)   The graph of $f(x) = \sqrt{x - 1}$ is obtained by shifting the graph of $y = \sqrt{x}$ one unit horizontally to the right.

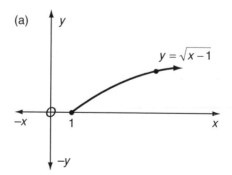

(b)   The graph of $f(x) = \sqrt{x} - 1$ is obtained by shifting the graph of $y = \sqrt{x}$ one unit vertically down.

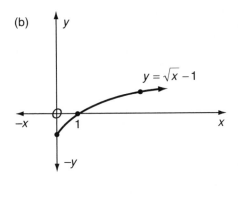

(b)

$y = \sqrt{x} - 1$

Graphing $y = a(x - h)^2 + k$:   Some graphs are combinations of horizontal and vertical shifts of familiar graphs. For example, if $g(x) = (x - 2)^2 + 1$, then $h = 2$ and $k = 1$. Hence, the graph of $g(x)$ can be obtained by shifting the graph of $y = x^2$ horizontally 2 units to the right and vertically 1 unit up. Since the turning point of $y = x^2$ is $(0, 0)$ the turning point of $g(x) = (x - 2)^2 + 1$ is $(0 + 2, 0 + 1) = (2, 1)$. In general, the graph of $y = a(x - h)^2 + k$ is the parabola $y = ax^2$ shifted so that its turning point is (h, k) and its axis of symmetry is $x = h$.

KEY EXAMPLE

Sketch the graph of $f(x) = |x + 2| + 1$

Solution: Since

$$f(x) = |x + 2| + 1 = |x - (-2)| + 1,$$

$h = -2$ and $k = 1$. Hence, the graph of $f(x)$ is obtained by shifting the graph of $y = |x|$ 2 units horizontally to the left and 1 unit vertically up.

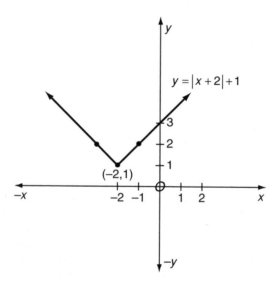

KEY EXAMPLE

Put the equation $y = 3x^2 + 6x - 5$ into the form $y = a(x - h)^2 + k$ and describe its graph.

Solution: After writing $y = 3(x^2 + 2x) - 5$, complete the square for the expression inside of the parentheses. The number that completes the square is obtained by taking one-half the coefficient of x and squaring it, which results in 1. Since this number is multiplied by 3, at the same time subtract 3 times this value outside the parentheses:

$$y = 3(x^2 + 2x + 1) - 3 \cdot 1 - 5 = 3(x + 1)^2 - 8$$

Since
$$y = 3(x + 1)^2 - 8$$
is equivalent to
$$y = 3[x - (-1)]^2 + (-8),$$

$h = -1$ and $k = -8$. Thus, the graph of $y = 3(x + 1)^2 - 8$ is the parabola $y = 3x^2$ shifted 1 unit horizontally to the left and 8 units vertically down so that its new turning point is $(-1, -8)$ and its axis of symmetry is $x = -1$.

Key 35 The composition of two
functions

OVERVIEW *Starting with two functions* f *and* g *we can
define a new function,* h, *by taking a function of a function
and writing* h(x) = g(f(x)). *Function* h *is called a **composite function**.*

Example of a composite function: If $f(x) = x - 5$ and $g(x) = \sqrt{x}$,
then $g(f(9))$ represents the function value that results when $x = 9$ is
the input value for function f, and the corresponding range value, $f(9)$,
becomes the input value of function g. Since $f(9) = 9 - 5 = 4$ and
$g(4) = \sqrt{4} = 2$, $g(f(9)) = g(4) = 2$.

Composition of functions f and g: The **composite of function f
followed by function g** is denoted by **$g \circ f$** (read as "g of f") and is
defined as the set of function values

$$(g \circ f)(x) = g(f(x))$$

for all x that is in the domain of f such that $f(x)$ is in the domain of
g. The notation $g(f(x))$ represents a particular function value and is
read as "g of f of x."

Order of composition matters: Functions f and g can be composed in
either order. The **composite of function g followed by function f** is
denoted by **$f \circ g$** and consists of the set of function values $f(g(x))$. In
the first example of a composite function where $f(x) = x - 5$ and
$g(x) = \sqrt{x}$, we saw that $g(f(9)) = 2$. Similarly, $f(g(9)) = f(3) = -2$.
Thus, $g(f(9)) \neq f(g(9))$. In general, the composition of functions is
not a commutative operation.

Domain of a composite function: A composite function, for example
$g \circ f$, is defined so that the domain of $g \circ f$ is restricted to the largest
possible set of real numbers for which x is in the domain of function
f and $f(x)$ is in the domain of function g. For example, if $f(x) = x - 5$
and $g(x) = \sqrt{x}$ $(x \geq 0)$, then the domain of the composite function
$g \circ f$ consists only of those values of x in the domain of function f
for which $g(f(x))$ is real. If x is less than 5, then $f(x)$ is negative
which means $g(f(x))$ is not real. Thus, if $x = 4$, then $f(4) = -1$ and
$g(f(4)) = \sqrt{-1}$ so $g(f(4))$ is not real. Hence, the domain of $g \circ f$
must be restricted to all numbers x that are greater than or equal to 5.

Finding an equation for a composite function: If $f(x) = x^2$ and
$g(x) = 3x + 1$, then an equation of the composite function $g \circ f$ can be

obtained by starting with $g(f(x))$ and replacing $f(x)$ with x^2, and then applying the function rule for $g(x)$. Thus,

$$g(f(x)) = g(x^2) = 3[x^2] + 1.$$

Similarly, the composite function $f \circ g$, can be derived by starting with $f(g(x))$ and replacing $g(x)$ with $3x + 1$, and then applying the function rule for $f(x)$:

$$f(g(x)) = f(3x + 1) = [3x + 1]^2 = 9x^2 + 6x + 1$$

Since $g(f(x))$ and $f(g(x))$ are defined by different equations, once again we see that the composition of two functions is *not* a commutative operation.

KEY EXAMPLE

If $f = \{(1, 3)(2, 4)\}$ and $g = \{(1, 2)(3, 6)\}$, find

  (a) $(g \circ f)(1)$ (b) $(f \circ g)(1)$ (c) $g(f(2))$

Solution: (a) $(g \circ f)(1) = g(f(1)) = g(3) = 6$

(b) $(f \circ g)(1) = f(g(1)) = f(2) = 4$

(c) $g(f(2)) = g(4)$. Since 4 is not in the domain of g, $g(f(2))$ is not defined.

KEY EXAMPLE

If $g(x) = \dfrac{x-1}{2}$ and $h(x) = 2x + 1$, find: (a) $h(g(x))$ (b) $h(h(3))$

Solutions: (a) $h(g(x)) = h\left(\dfrac{x-1}{2}\right) = 2\left(\dfrac{x-1}{2}\right) + 1 = x.$

(b) Since $h(h(x)) = h(2x) = 2(2x) = 4x$, $h(h(3)) = 4 \cdot 3 = 12$.

KEY EXAMPLE

If $g(x) = \sqrt{x-2}$ and $h(x) = \left(\dfrac{1}{x}\right)^2$, find $h(g(x))$ and state the domains of functions g and $h \circ g$.

Solution: For $h(g(x)) = \dfrac{1}{x-2}$, $x \neq 2$. Since the domain of $g(x)$ is $x \geq 2$ and $x \neq 2$, the domain of $h \circ g$ is $x > 2$.

Key 36   Inverse functions

OVERVIEW   *For some functions there exist other functions that can reverse their effect. Pairs of functions that are related in this way are called **inverse functions**.*

Example of inverse functions: Functions like $f(x) = x^2$ and $g(x) = \sqrt{x}$ are inverse functions since they undo the effect of each other. For example, if $x = 3$, then $f(3) = 9$ and $g(9) = \sqrt{9} = 3$ which is the value of x we started with. Thus, $g(f(3)) = 3$. Also, $f(g(3)) = 3$. Notice that the ordered pair (3, 9) belongs to function f and the corresponding ordered pair for function g is (9, 3). If functions f and g are inverse functions, then the domain of f is the range of g, and the range of f is the domain of g.

Definition of inverse functions: Functions f and g are inverse functions if and only if

Equation 1 $\qquad\qquad g(f(x)) = x$

and

Equation 2 $\qquad\qquad f(g(x)) = x$

provided that in equation 1, x is in the domain of f and $f(x)$ is in the domain of g, and in equation 2, x is in the domain of g and $g(x)$ is in the domain of f. Thus, functions f and g are inverse functions if their composition, taken in either order, always produces the original input value.

Forming inverse functions: To find the inverse of a function, if it exists, we need to interchange its domain and range.

- If the function is defined by a set of ordered pairs, then its inverse is obtained by swapping the positions of the first and second members of each ordered pair. For example, if $f = \{(-1, 0), (0, 1), (1, 2)\}$, then the inverse of f is $\{(0, -1), (1, 0), (2, 1)\}$.
- If the function is defined by an equation, then its inverse is formed by interchanging x and y, and then solving the resulting equation for y. If $f(x) = 2x + 3$, let $y = 2x + 3$. Interchanging x and y yields $x = 2y + 3$. Solving this equation for y gives the inverse of $f(x)$, $y = \dfrac{x-3}{2}$.

Inverse functions: Not all functions have inverse functions. Interchanging the x and y values of the function $f = \{(-2, 4), (2, 4)\}$ gives $\{(4, 2), (4, -2)\}$ which is not a function. If f is a one-to-one function (*f* passes the horizontal line test), then its inverse is a function and is denoted by f^{-1}. Be careful not to interpret the -1 in f^{-1} as an exponent. The notation f^{-1} is read, "*f*–inverse."

Making functions one-to-one: Sometimes the domain of a function must be restricted so that its inverse can be obtained. For example, $f(x) = x^2$ is not a one-to-one function. By restricting the domain of *f* to the set of nonnegative real numbers ($x \geq 0$), we insure *f* has an inverse, namely $f^{-1}(x) = \sqrt{x}$.

Graphs of inverse functions: For every point (a, b) contained on the graph of a one-to-one function, the point (b, a) is contained on the graph of its inverse. As shown in the accompanying diagram, this means that the graph of a function and its inverse are symmetric with respect to the line $y = x$. Either graph may be obtained by reflecting the other graph in the line $y = x$.

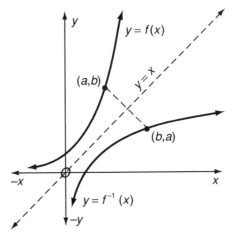

KEY EXAMPLE

If $h(x) = 1 - x^3$, find $h^{-1}(9)$.

Solution: Find h^{-1} by letting $y = 1 - x^3$ and interchanging *x* and *y*, obtaining $x = 1 - y^3$. Solving for *y* gives

$$y = h^{-1}(x) = \sqrt[3]{1 - x}.$$

Hence,
$$h^{-1}(9) = \sqrt[3]{1 - 9} = \sqrt[3]{8} - = 2.$$

KEY EXAMPLE

If $f(x) = 1.9x^3 - 17$, find $f(f^{-1}(6.4))$.

Solution: Since *f* and f^{-1} are inverse functions, $f(f^{-1}(x)) = x$. Hence,

$$f(f^{-1}(6.4)) = 6.4.$$

Key 37 Polynomial and rational functions

OVERVIEW *A polynomial function in* x *contains only positive integer powers of* x. *A rational function in* x *is the quotient of two polynomial functions in* x.

Polynomial function of degree n: A polynomial function $P(x)$ of degree n has the general form

$$P(x) = a_n x^n + a_{n-1} x^{n-1} + a_{n-2} x^{n-2} + \ldots a_1 x + a_0$$

where $a_n \neq 0$, n is a positive integer, and a_n, a_{n-1}, a_{n-2}, $\ldots a_1$, a_0 are constants. The coefficient of the greatest power of x, a_n, is called the **leading coefficient**. In the polynomial function $f(x) = 2x^4 - x^2 + 7x - 5$, the leading coefficient is 2. Since the term x^3 is missing, its coefficient is 0. Thus, $a_4 = 2$, $a_3 = 0$, $a_2 = -1$, $a_1 = 7$, and $a_0 = -5$.

Definition of rational function: If $P(x)$ and $Q(x)$ are polynomial functions, then $f(x) = \dfrac{P(x)}{Q(x)}$ is a rational function the domain of which excludes those values of x, if any, for which $Q(x) = 0$.

Point discontinuities:

If $f(x) = \dfrac{P(x)}{Q(x)}$ and

$P(x)$ and $Q(x)$ have $x - c$ as a common factor, then the graph of $f(x)$ has a point discontinuity or "hole" at $x = c$. For example, since

$$f(x) = \frac{x^2 - 4}{x - 2}$$

$$= \frac{(x-2)(x+2)}{x-2}$$

$$= x + 2$$

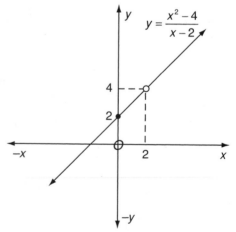

Figure 6.12 Graph of $f(x) = \dfrac{x^2 - 4}{x - 2}$

the graph of $f(x)$ has a point discontinuity at $x = 2$. Thus, the graph of $y = f(x)$ is identical to the graph of $y = x + 2$ except that there is an open circle at $x = 2$ to indicate that graph does not contain this point. See Figure 6.12.

Asymptote: An asymptote is a line that a curve approaches, but does not intersect as the curve is extended indefinitely. A curve may have vertical asymptotes, horizontal asymptotes, or both types of asymptotes.

- A line $x = a$ is **vertical asymptote** of the graph of $y = f(x)$ if as $x \to a$ (read "x approaches a"), $f(x) \to \infty$ or $f(x) \to -\infty$ (read "f of x approaches infinity" or "f of x approaches negative infinity"). The expression "approaches infinity" means increases without having an upper bound while the expression "approaches negative infinity" means decreases without having a lower bound. In Figure 6.13 the y-axis (the line $x = 0$) is a vertical asymptote of the graph of $f(x) = \dfrac{1}{x^2}$ since as x approaches 0 from either the positive or the negative side of 0, the graph of $f(x)$ constantly rises while getting closer and closer to the y-axis without ever touching it.

- A line $y = b$ is a **horizontal asymptote** of the graph of $y = f(x)$ if, as $x \to \infty$ or as $x \to -\infty$, $f(x) \to b$. In Figure 6.13, the graph of $f(x) = \dfrac{1}{x^2}$, has the x-axis (the line $y = 0$) has a horizontal asymptote since as x increases (or decreases) without bound, the graph of $f(x)$ constantly falls while getting closer and closer to the x-axis without ever touching it.

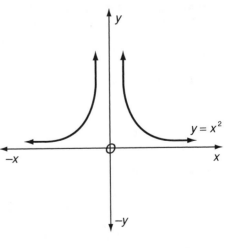

Figure 6.13 Graph of $f(x) = \dfrac{1}{x^2}$

Key 38   Graphing rational functions

OVERVIEW   *A rough sketch of a rational function can be made if you know some key features like intercepts, asymptotes, and symmetry properties.*

Finding intercepts of $y = P(x)/Q(x)$: The x-intercepts are the real solutions, if any, of the equation $P(x) = 0$, provided these values of x do not make $Q(x)$ equal to 0 at the same time. The y-intercept, if any, is obtained by letting $x = 0$ and evaluating $y = \dfrac{P(0)}{Q(0)}$, provided the quotient is defined.

Finding vertical asymptotes of $y = P(x)/Q(x)$: The line $x = a$ is a vertical asymptote if $Q(a) = 0$, but $P(a) \neq 0$. Thus, a vertical asymptote is found at each value of x, if any, for which the denominator is 0 and the numerator is not 0.

Example: Any vertical asymptotes of $y = \dfrac{4x + 8}{x^2 - x - 6}$ can be determined by writing y in factored form as $y = \dfrac{4(x + 2)}{(x - 3)(x + 2)}$. Although $x = 3$ and $x = -2$ both make the denominator 0, only $x = 3$ is a vertical asymptote. Since $x = -2$ makes the numerator and the denominator 0 at the same time, the graph will have a discontinuity at this point.

Finding a horizontal asymptote of $y = P(x)/Q(x)$: Suppose $P(x)$ is an nth degree polynomial and $Q(x)$ is an mth degree polynomial. The graph of $f(x) = P(x)/Q(x)$ has a horizontal asymptote when $n \leq m$, and no horizontal asymptote when $n > m$.

- If $n < m$, then the x-axis is a horizontal asymptote. For example, the graph of $f(x) = \dfrac{4x}{x^2 - 1}$ has the x-axis as a horizontal asymptote since the degree of the numerator is less than the degree of the denominator.

- If $n = m$, then the line $y = \dfrac{a}{b}$ is a horizontal asymptote where a is the leading coefficient of $P(x)$ and b is the leading coefficient of $Q(x)$. For example, the graph of $f(x)$ $\dfrac{2x^2 + x + 4}{3x^2 - 5x + 6}$ has the line $y = \dfrac{2}{3}$ as a horizontal asymptote.

Finding a slant asymptote of $y = P(x)/Q(x)$: If the degree of the numerator of a rational function exceeds the degree of its denominator by 1, then its graph has a **slant** (oblique) asymptote. An equation of any oblique asymptote is obtained by dividing the numerator by the denominator, deleting any remainder, and then setting y equal to the remaining part of the quotient. For example, the graph of

$$y = \frac{3x^2 - 12x + 2}{x - 4}$$ has an oblique asymptote. Dividing the numerator

by the denominator using long division yields $3x + \dfrac{2}{x - 4}$.

Discarding the fractional term which represents the remainder and setting y equal to the remaining part of the quotient gives $y = 3x$ as the equation of an oblique asymptote.

Other special features: Sketching a rational function can sometimes be simplified by recognizing the following:

- The graph has symmetry. Rational functions, like polynomial functions, may be symmetric to a line or to the origin. For example, $f(x) = \dfrac{1}{x^2}$ is an even function since $f(-x) = f(x)$ so its graph has the y-axis as a line of symmetry, as shown in Figure 6.13 in Key 37.

- The graph may be obtained by shifting a familiar graph. For example, the graph of $F(x) = \dfrac{1}{(x-2)^2}$ can be obtained by shifting the graph of $f(x) = \dfrac{1}{x^2}$ two units to the right so $x = 2$ is a vertical asymptote.

KEY EXAMPLE

Sketch the graph of $f(x) = \dfrac{x^2 - 36}{x^2 - 9}$.

Solution: Factoring $f(x)$ gives $f(x) = \dfrac{(x-6)(x+6)}{(x-3)(x+3)}$.

x-intercepts: $(-6, 0)$ and $(6, 0)$ since if $y = 0$, $x^2 - 36 = 0$ so $x = \pm 6$.

y-intercept: $(0, 4)$ since if $x = 0$, $y = f(0) = \dfrac{-36}{-9} = 4$.

Vertical asymptotes: The lines $x = -3$ and $x = 3$ since ± 3 make the denominator 0, but not the numerator.

Horizontal asymptote: The line $y = 1$ since the numerator and denominator have the same degree and the ratio of their leading coefficients is 1.

Additional points: Plot at least one point on either side of any asymptote.

x	-4	-2	4
$f(x)$	-3.86	6.4	-3.86

Line of symmetry: The y-axis since $f(-x) = f(x)$.

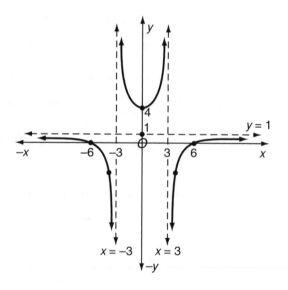

108

Key 39 Equations that describe variation

OVERVIEW *Sometimes two or more variables are related by the operations of multiplication or division or by a combination of these operations. These relationships can be described as **direct**, **inverse**, or **joint variation**.*

Direct variation: If variables x and y are related so that the ratio of y to x always remains the same, then we say y *varies directly as x* or y *is directly proportional to x*. Expressed in symbols,

$$\frac{y}{x} = k \quad \text{or} \quad y = kx$$

where k stands for a number called the **constant of variation**. Multiplying x or y by a fixed amount always causes the other variable to be *multiplied* by the same fixed amount. The graph of a direct variation is a line through the origin whose slope is the constant of variation.

Inverse variation: If variables x and y are related so that their product always remains the same, then we say that y *varies inversely as* x *or* y *is inversely proportional to* x. Expressed in symbols,

$$xy = k \quad or \quad y = \frac{k}{x}$$

where k is a nonzero number called the **constant of variation**. Multiplying x or y by some fixed amount always caused the other variable to be *divided* by the same fixed amount. The graph of an inverse variation is an equilateral hyperbola.

Variation involving powers: A variable may vary directly or indirectly with a power. For some nonzero constant k,

- If y varies directly as the nth power of x, then $y = k\,x^n$
- If y varies indirectly as the nth power of x, then $y\,x^n = k$

KEY EXAMPLE

If p varies directly as the square of q, and $p = 147$ when $q = 7$, find the value of q when $p = 75$.

Solution: Since p varies directly as the square of q, $p = k\,q^2$. Solve for k by letting $p = 147$ and $q = 7$,

$$147 = k\,(7^2)$$

so

$$k = \frac{147}{49} = 3.$$

Using $p = 3 q^2$ as the equation of variation, let $p = 75$ and solve for q,

$$75 = 3 q^2$$

so

$$q^2 = 25 \text{ and } q = \pm 5.$$

KEY EXAMPLE

If y varies inversely as x, and $y = 33$ when $x = 6$, find y when $x = 9$.

Solution: If y varies inversely as x, then $xy = k$. Letting $x = 6$ and $y = 33$ gives $k = (6)(33) = 198$. Using the equation of variation $xy = 198$, let $x = 9$ and solve for y,

$$9y = 198$$

so

$$y = \frac{198}{9} = 22.$$

KEY EXAMPLE

Find an equation of variation when s varies inversely as the square of t, and $s = 2$ when $t = 4$.

Solution: Since $st^2 = k$, $2 (4^2) = k$ so $k = 32$. Thus, $st^2 = 32$ is an equation of variation.

Joint variation: If **z varies jointly as x and y**, then $z = kxy$, where k is some nonzero constant. Joint variation, direct variation, and inverse variation may occur in various combinations and may involve powers as well as roots of variables.

Example 1: If z varies jointly as x and the cube of y, then $z = k x y^3$.

Example 2: If z varies directly as the square of x and inversely as the product of w and the cube root of y, then

$$z = k \frac{x^2}{w\sqrt[3]{y}} .$$

KEY EXAMPLE

The quantity z varies jointly as x and the square of y. If $z = 2$ when $x = 4$ and $y = 3$, find y when $z = 10$ and $x = 5$.

Solution: Since $z = k\,xy^2$, $2 = k(4 \cdot 3^2)$ so $k = \dfrac{2}{36} = \dfrac{1}{18}$. Hence, $z = \dfrac{1}{18}\,xy^2$. If $z = 10$ and $x = 5$, then $10 = \dfrac{5y^2}{18}$. Solving for y^2 leads to $y^2 = \dfrac{180}{5} = 36$, so $y = \pm 6$.

Theme 7 POLYNOMIAL EQUATIONS

*T*he degree of the polynomial equation $P(x) = 0$ is the degree of the polynomial $P(x)$. Some polynomial equations of degree greater than 2 can be solved by factoring, as in

$$3x^3 + x^2 - 12x - 4 = 0$$

$$x^2(3x + 1) - 4(3x + 1) = 0$$

$$(x^2 - 4)(3x + 1) = 0$$

If $x^2 - 4 = 0$, then $x = \pm 2$. If $3x + 1 = 0$, then $x = \dfrac{-1}{3}$.

Other polynomial equations cannot be factored as easily or cannot be factored at all. If some of the roots of a polynomial equation of degree greater than 2 are known, then it may be possible to use this information to reduce the degree of the original equation so that a quadratic equation eventually results.

	INDIVIDUAL KEYS IN THIS THEME
40	Synthetic division
41	The remainder and factor theorems
42	Zeros of polynomial functions
43	Reducing the degree of a polynomial equation
44	The rational root theorem
45	Descartes' rule of signs
46	Locating and approximating irrational roots

Key 40 Synthetic division

OVERVIEW *A condensed form of the long division process, called* **synthetic division**, *can be used whenever a polynomial* P(x) *is divided by a first degree binomial of the form* x − r *where* r *is some constant.*

Synthetic division: To divide $P(x) = 4x^3 - 7x^2 - 16x + 5$ by $x - 3$ using synthetic division, write the synthetic divisor 3 followed by the coefficients of the powers of x using the format illustrated below:

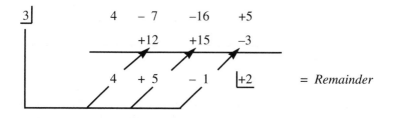

Bring down the first detached coefficient, 4, as shown above, and multiply it by the synthetic divisor, 3. Write the product of +12 above the horizontal line in the next column. Repeat this process as illustrated in the table that follows:

$$
\begin{array}{r|rrrr}
3 & 4 & -7 & -16 & +5 \\
  &   & +12 & +15 & -3 \\
\hline
  & 4 & +5 & -1 & \boxed{+2} \quad = Remainder
\end{array}
$$

Reading the last line of the synthetic division table from left to right gives the detached coefficients of the second degree polynomial quotient in standard form and the constant remainder. The quotient is $4x^2 + 5x - 1$ with remainder 2. Thus,

$$
\frac{4x^3 - 7x^2 - 16x + 5}{x - 3} = 4x^2 + 5x - 1 + \frac{2}{x - 3}.
$$

Guidelines for synthetic division: In dividing the polynomial $P(x)$ synthetically,

- Arrange the terms of $P(x)$ in standard form. If a power of x is missing from $P(x)$, then a coefficient of 0 is used in that position.
- If the divisor has the form $x - r$, use r as the synthetic divisor; if the divisor has the form $x + r$, the synthetic divisor is $-r$.
- The quotient obtained will always be a polynomial of degree one less than the degree of $P(x)$.

KEY EXAMPLE

Divide $x^4 + 2x^3 - 5x + 1$ by $x + 3$.

Solution: Provide for the missing power of x in the dividend by writing $x^4 + 2x^3 + 0\,x^2 - 5x + 1$. Put the divisor in the form $x - r$ by writing $x - (-3)$ so the synthetic divisor is -3.

$$
\begin{array}{r|rrrrr}
-3 & 1 & +2 & +0 & -5 & +1 \\
   &   & -3 & +3 & -9 & \underline{+42} \\
\hline
   & 1 & -1 & +3 & -14 & \underline{+43} \quad = \ Remainder \\
\end{array}
$$

$$Quotient = \quad x^3 \quad -x^2 \quad +3x \quad -14$$

Thus,

$$\frac{x^4 + 2x^3 - 5x + 1}{x + 3} = x^3 - x^2 + 3x - 14 + \frac{43}{x + 3}$$

KEY EXAMPLE

Divide $4x^3 - 8x^2 + x + 11$ by $2x - 3$.

Solution: Since the divisor can be written as $2(x - 3/2)$, divide $4x^3 - 8x^2 + x + 11$ synthetically by $x - 3/2$ and then divide the quotient by 2:

$$
\begin{array}{r|rrrr}
\frac{3}{2} & 4 & -8 & +1 & +11 \\
   &   & +6 & -3 & -3 \\
\hline
   & 4 & -2 & -2 & \underline{+8} \quad = \ Remainder \\
\end{array}
$$

Dividing the detached coefficients of the quotient by 2 gives a quotient of $2x^2 - x - 1$ with a remainder of 8. Hence,

$$\frac{4x^3 - 8x^2 + x + 11}{2x - 3} = 2x^2 - x - 1 + \frac{8}{2x - 3}$$

KEY EXAMPLE

If $x - 2$ is a factor of $P(x) = x^3 + 2x^2 - 5x - 6$, factor $P(x)$ completely.

Solution: Divide $P(x)$ by $x - 2$ synthetically to obtain the corresponding quadratic factor of $P(x)$.

$$
\begin{array}{r|rrrr}
2 & 1 & +2 & -5 & -6 \\
  &   & +2 & +8 & +6 \\
\hline
  & 1 & +4 & +3 & \underline{+0}
\end{array}
$$

Since the quotient is $x^2 + 4x + 3$,

$$x^3 + 2x^2 - 5x - 6 = (x - 2)(x^2 + 4x + 3)$$

Factor $(x^2 + 4x + 3)$: $\qquad = (x - 2)(x + 1)(x + 3).$

Key 41 The remainder and factor theorems

OVERVIEW *The remainder obtained in dividing* P(x) *by* x − r *synthetically also can be obtained by evaluating* P(r). *If* P(r) = 0, *then* P(x) *is evenly divisible by* x − r *so* x − r *is a factor of* P(x).

The remainder theorem: If a polynomial $P(x)$ is divided by $x - r$, then the remainder is equal to the function value $P(r)$. In Key 40 we divided $P(x) = 4x^3 - 7x^2 - 16x + 5$ by $x - 3$ and obtained 2 as the remainder. This remainder, but not the quotient, may also be obtained by evaluating $P(3)$. Since $P(3) = 4(3^3) - 7(3^2) - 16(3) + 5 = 2$, the remainder is 2.

The factor theorem: A polynomial $P(x)$ has $x - r$ as a factor if and only if $P(r) = 0$. For example, to show that $x + 1$ is a factor of $P(x) = 2x^3 - 9x^2 + x + 12$, write $x + 1$ as $x - (-1)$ and evaluate $P(-1)$. Since $P(-1) = 2(-1)^3 - 9(-1)^2 + (-1) + 12 = 0$, $x + 1$ is a factor of $P(x)$. To find the polynomial that is the corresponding factor of $x + 1$, we would need to divide $P(x)$ by $x + 1$, as was done in the example in Key 40 (page 113).

KEY EXAMPLE

For what value of k is $(x + 3)$ a factor of $x^4 + x^3 - x^2 + kx - 12$?

Solution: The binomial $x + 3$ can be written as $x - (-3)$. Thus, $x - (-3)$ is a factor of $P(x) = x^4 + x^3 - x^2 + kx - 12$ if $P(-3) = 0$. Replacing x with -3 in $P(x)$ and setting the result equal to 0 gives

$$(-3)^4 + (-3)^3 - (-3)^2 + k(-3) - 12 = 0$$
$$81 - 27 - 9 - 3k - 12 = 0$$
$$33 - 3k = 0$$
$$k = \frac{-33}{-3} = 11$$

Thus, $(x + 3)$ is a factor of $x^4 + x^3 - x^2 + 11x - 12$.

Key 42   Zeros of polynomial functions

OVERVIEW   *Any value of* x *for which a polynomial* P(x) *equals 0 is a **zero** of the function and a **root** of the polynomial equation,* P(x) = 0.

Zero of a polynomial function *P(x)*: A zero of a polynomial function is any number r, real or complex, for which $P(r) = 0$. If any one of the following statements is true, then so are each of the other statements:

- r is a zero of the polynomial function $P(x)$;
- $x = r$ is a root of the polynomial equation $P(x) = 0$;
- $x - r$ is a factor of the polynomial $P(x)$;
- r is an x-intercept of the graph of $y = P(x)$, provided r is a real number.

Examples:

1. If $P(x) = x^2 - 6x + 9$, then 3 is a zero of $P(x)$ since $P(3) = 0$. Also, 3 is a root of the equation $x^2 - 6x + 9 = 0$ and, consequently, $x - 3$ is a factor of $x^2 - 6x + 9$.
2. To find the zeros of $P(x) = x^3 - 4x$, find the roots of the equation $x^3 - 4x = 0$. Since $x^3 - 4x = x(x - 2)(x + 2) = 0$, the zeros of $P(x)$ are 0 and ±2. Hence, the x-intercepts of the graph of $y = x^3 - 4x$ occur at $x = -2$, $x = 0$, and $x = 2$.

Fundamental theorem of algebra: Every nonconstant polynomial function has at least one complex zero.

Polynomial factorization theorem: A nonconstant polynomial $P(x)$ of degree n can be factored into n linear binomial factors. Thus, there exist numbers $r_1, r_2, r_3, \ldots r_n$ real or complex, such that

$$P(x) = c(x - r_1)(x - r_2)(x - r_3) \ldots (x - r_n)$$

where c is a nonzero constant. The numbers $r_1, r_2, r_3, \ldots, r_n$ are the zeros of $P(x)$.

Multiplicity of zeros: The polynomial factorization theorem does not guarantee that the numbers $r_1, r_2, r_3, \ldots, r_n$ are all different. If the binomial $(x - r)$ appears as a factor k times, then r is said to be a **zero (root) of multiplicity *k***. For example, if

$$P(x) = (x - 3)^2 (x + 1)^3 (x - 5),$$

then 3 is a zero (or root) of *multiplicity two* since $x - 3$ appears as a factor *two* times. Thus, -1 is a zero (or root) of *multiplicity three*. The polynomial $P(x)$ has 5 as a **simple zero** since its multiplicity is one.

n-Roots theorem of polynomial equations: A polynomial equation of degree n has exactly n roots, real or complex, provided that a root of multiplicity k is counted k times.

Although its roots may be difficult to find, the *fourth* degree polynomial equation $x^4 + x^3 - 5x^2 + 2x + 7 = 0$ has a total of *four* roots. The fifth degree equation $(x^2 + 1)(x - 6)^3 = 0$ has a total of *five* roots. Two of these roots are the imaginary numbers $\pm i$ which correspond to the roots of the equation $x^2 + 1 = 0$. Although the equation has only one distinct real root, namely 6, it is a root of multiplicity three so it is counted three times.

Roots that occur in pairs: Nonreal solutions of polynomial equations with real coefficients occur in conjugate pairs. That is, if $a + bi$ is a root of a polynomial equation (where $i = \sqrt{-1}$), then $a - bi$ is also a root. If a polynomial equation with rational coefficients has irrational roots, then they also occur in conjugate pairs. That is, if $a + \sqrt{b}$ is a root of the equation, then $a - \sqrt{b}$ is also a root.

KEY EXAMPLE

Find the polynomial equation $P(x) = 0$ of lowest degree and with rational coefficients if its solution set includes $2i$, $-\sqrt{5}$, and 3 as a root of multiplicity 2.

Solution: For each root r of $P(x) = 0$, determine the corresponding factor $x - r$. If $2i$ is a root of $P(x) = 0$, then its conjugate, $-2i$, is also a root. Hence, $(x - 2i)$ and $(x + 2i)$ are factors of $P(x)$. Similarly, $(x + \sqrt{5})$ and $(x - \sqrt{5})$ are also factors of $P(x)$. Since 3 is a root of multiplicity 2, $(x - 3)^2$ is also a factor of $P(x)$. Thus, the equation in factored form is

$$(x - 2i)\,(x + 2i)\,(x + \sqrt{5})\,(x - \sqrt{5})\,(x - 3)^2 = 0.$$

Multiplying the factors together gives the sixth degree equation,

$$x^6 - 6x^5 + 8x^4 + 6x^3 - 29x^2 + 120x - 180 = 0.$$

Key 43 Reducing the degree of a polynomial equation

OVERVIEW *The degree of a polynomial equation is reduced by one each time the equation is divided by a factor that corresponds to a root. By removing a sufficient number of roots from a polynomial equation, it may be possible to obtain a quadratic equation.*

Depressed equation: If $x - r$ is a factor of an nth degree polynomial $P(x)$, then the equation obtained by dividing $P(x)$ by $x - r$ and setting the quotient equal to 0 is an equation of degree $n - 1$ called the **depressed equation**. If we know one or more roots of a polynomial equation, it may be possible to obtain a second degree depressed equation which can be solved either by factoring or, if necessary, by using the quadratic formula.

KEY EXAMPLE

Given the equation $P(x) = x^3 - 13x + 12 = 0$, find one root by inspection and the other roots by solving the depressed equation.

Solution: By inspecting the equation $P(x) = x^3 - 13x + 12 = 0$, we see that $P(1) = 0$, so $x = 1$ is a solution. The depressed equation is obtained by dividing $x^3 - 13x + 12$ by $x - 1$ synthetically and setting the quotient equal to 0. This gives $x^2 + x - 12 = 0$ whose roots are 3 and –4.

Thus, the three roots of the original equation are 1, 3, and –4.

KEY EXAMPLE

If two roots of the equation $P(x) = x^4 + x^3 - 15x^2 - 22x + 8 = 0$ are –2 and 4, find the remaining roots.

Solution: Divide $P(x)$ by $x - (-2)$ and then divide the quotient obtained by $x - 4$.

$$
\begin{array}{r|rrrrr}
-2\,] & 1 & 1 & -15 & -22 & +8 \\
& \downarrow & -2 & +2 & -26 & -8 \\
\hline
4\,] & 1 & -1 & -13 & +4 & \,\rfloor\,0 \\
& \downarrow & +4 & +12 & -4 & \\
\hline
& 1 & +3 & -1 & \,\rfloor\,0 &
\end{array}
$$

Quotient: $x^2 + 3x - 1$

Depressed Equation: $x^2 + 3x - 1 = 0$

The roots of the depressed equation can be obtained using the quadratic formula:

$$x = \frac{-3 \pm \sqrt{9 - 4(1)(-1)}}{2} = \frac{-3 \pm \sqrt{13}}{2}$$

Hence, the four roots of the original equation are -2, 4, and $\dfrac{-3 \pm \sqrt{13}}{2}$.

Key 44 The rational root theorem

OVERVIEW *The leading coefficient and the constant term of a polynomial equation provide clues to any possible rational roots of the equation.*

Integer root theorem: If an integer p is a root of a polynomial equation having integer coefficients and a leading coefficient of 1, then p must be a factor of the constant term of the equation. Since the equation $x^4 + x^3 - 15x^2 - 22x + 8 = 0$ has integer coefficients and its leading coefficient is 1, any integer roots must be factors of 8. Thus, the possible integer roots of this equation are limited to ± 1, ± 2, ± 4, and ± 8.

Rational root theorem: If p/q is a rational root in lowest terms of the polynomial equation with integer coefficients

$$a_n x^n + a_{n-1} x^{n-1} + a_{n-2} x^{n-2} + \ldots + a_1 x + a_0 = 0 \quad \text{(where } a_n \neq 0\text{)}$$

then p is a factor of a_0 and q is a factor of a_n.

Example. The equation $P(x) = 3x^4 - 7x^3 + 8x^2 - 14x + 4 = 0$ may or may not have rational roots. If the equation does have rational roots, then each rational root must be a fraction whose numerator is a factor of 4 and whose denominator is a factor of 3. Since the factors of 4 are ± 1, ± 2, and ± 4, and the factors of 3 are ± 1 and ± 3, the possible rational roots of the equation are limited to

$$\pm \frac{1}{3}, \ \pm \frac{2}{3}, \ \pm 1, \ \pm \frac{4}{3}, \ \pm 2, \text{ and } \pm 4.$$

Using synthetic division or the remainder theorem, we can systematically test each possible rational root to see which of these, if any, are actually roots of the given equation. Since $P(2) = 0$ and $P(1/3) = 0$, 2 and 1/3 are rational roots of the equation. Successively dividing $P(x)$ by the binomial factors that correspond to each root gives a second degree depressed equation:

$$\underline{1/3}\rfloor \quad 3 \quad -7 \quad 8 \quad -14 \quad +4$$

$$\downarrow \quad +1 \quad -2 \quad +2 \quad -4$$

$$\underline{2}\rfloor \quad 3 \quad -6 \quad +6 \quad -12 \quad \lfloor 0$$

$$\downarrow \quad +6 \quad +02 \quad +12$$

$$3 \quad 0 \quad +6 \quad \lfloor 0$$

The last depressed equation is $3x^2 + 6 = 0$ whose roots are $\pm i\sqrt{2}$. Hence, the four roots of the original equation are 1/3, 2, and $\pm i\sqrt{2}$.

Key 45   Descartes' rule of signs

OVERVIEW   *Descartes' rule of signs gives the possible combinations of the different types of roots—positive, negative, and nonreal—that a polynomial equation may have. Using this information, the number of possible rational roots that need to be tested can sometimes be reduced.*

Variation in sign: A variation in sign occurs when the numerical coefficients of two consecutive terms of a polynomial in standard form have opposite signs. Reading from left to right, the polynomial $x^4 - 2x^2 - 4x + 5$ has two variations in sign since the signs of the numerical coefficients of the terms, are

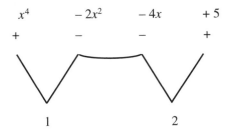

Descartes' rule of signs: This rule, named for the philosopher and mathematician René Descartes, relates the type of roots the polynomial equation $P(x) = 0$ may have to the number of variations in signs of $P(x)$.

- The number of positive real roots of a polynomial equation $P(x) = 0$ is equal to the number of variations in sign of $P(x)$ or that number reduced by an even number. The equation $x^3 - 6x^2 + 5x - 3 = 0$ has three variations in sign so it has either 3 real positive roots or 1 real positive root since 3 reduced by 2 is 1.

- The number of negative real roots of a polynomial equation $P(x) = 0$ is equal to the number of variations in sign of $P(-x)$ or that number reduced by an even number. For the equation $P(x) = x^3 - 6x^2 + 5x - 3 = 0$,

$$
\begin{aligned}
P(-x) &= (-x)^3 - 6(-x)^2 + 5x - 3 \\
&= -x^3 - 6x^2 + 5x - 3
\end{aligned}
$$

Since $P(-x)$ has two variations in sign, the original equation has either two negative real roots or no negative real roots since 2 reduced by 2 is 0. The possible combinations of the different types of roots that the equation $x^3 - 6x^2 + 5x - 3 = 0$ may have can be conveniently summarized using the table that follows:

Possible Root Combinations for the Equation $x^3 - 6x^2 + 5x - 3 = 0$

Possible Number of Positive Real Roots	Possible Number of Negative Real Roots	Number of Nonreal Roots
3	0	0
1	2	0
1	0	2

In completing this table keep in mind that nonreal roots, if any, must occur in pairs, and the total number of roots on each line of the table must equal the degree of the polynomial.

KEY EXAMPLE

Describe the types of roots that are possible for the equation

$$P(x) = 3x^4 - 7x^3 + 8x^2 - 14x + 4 = 0.$$

Solution: $P(x)$ has four variations in sign and $P(-x) = 3x^4 + 7x^3 + 8x^2 + 14x + 4 = 0$ has 0 variations in sign. The possible combinations of the different types of roots that the equation may have are summarized in the table that follows:

Possible Root Combinations for the Equation
$P(x) = 3x^4 - 7x^3 + 8x^2 - 14x + 4 = 0$

Possible Number of Positive Real Roots	Possible Number of Negative Real Roots	Number of Nonreal Roots
4	0	0
2	0	2
0	0	4

In Key 44 we saw that the equation $3x^4 - 7x^3 + 8x^2 - 14x + 4 = 0$ has six pairs of positive and negative numbers that are possible rational roots. If Descartes' rule of signs had been applied in the solution process, then we would have been able to shorten the list of possible rational roots by eliminating from consideration the six possible *negative* rational roots.

Key 46 Locating and approximating

irrational roots

OVERVIEW   *If a polynomial function* P(x) *changes sign in the interval* (a, b), *then a real root of the equation* P(x) = 0 *is located between* x = a *and* x = b. *If this root is irrational, then it can be approximated by "shrinking" the interval in which the root is located.*

Intermediate value theorem: If $P(x)$ is a polynomial with real coefficients and $P(a)$ and $P(b)$ have different signs, then the graph of $y = P(x)$ is a continuous curve that must cross the x-axis between $x = a$ and $x = b$. Thus, a real root of $P(x) = 0$ lies in the interval (a, b). For example, if $P(x) = x^3 + x - 7$, then $P(1) = -5$ and $P(2) = 3$. Since $P(1)$ and $P(2)$ have opposite signs, the equation $x^3 + x - 7 = 0$ has a real root between 1 and 2.

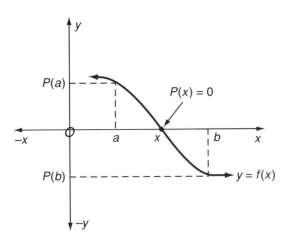

Bisection method for locating irrational roots: If we know the endpoints of an interval that contains an irrational root of $P(x) = 0$, then we can use the intermediate value theorem to find the half of the interval that contains this root. Repeating this process over and over "traps" the root in subintervals whose endpoints become closer and closer in value. This allows the root to be estimated to any desired accuracy.

KEY EXAMPLE

Approximate to the nearest tenth the positive irrational root of the equation $P(x) = x^3 + x - 7 = 0$.

Solution: Step 1. *Find two consecutive integers between which an irrational root lies.* Since $P(1) = -5$ and $P(2) = 3$, the equation has a real root between 1 and 2. The rational root theorem tells us that the only possible positive rational roots of the equation are 1 and 7. Hence, an irrational root is contained in the interval $(1, 2)$.

Step 2. *Determine the half of the interval that contains the root, and then discard the other half-interval.* The midpoint of the interval $(1, 2)$ is the average of the endpoints, 1.5. Since

$$P(1.5) = (1.5)^3 + 1.5 - 7 = -2.125 < 0$$

and

$$P(2) > 0,$$

the intermediate value theorem tells us that the root must be located in the subinterval $(1.5, 2)$.

Repeating Step 2 using the interval $(1.5, 2)$ locates the root in the subinterval $(1.5, 1.75)$. After repeating this process twice we find that root is contained in the interval $(1.6875, 1.75)$. Hence, the root correct to the nearest tenth is 1.7.

Theme 8 EXPONENTIAL AND LOGARITHMIC FUNCTIONS

*M*ost of our work with functions has been limited to *algebraic functions*. **Algebraic functions** are functions that can be defined in terms of the sums, differences, products, quotients, or roots of polynomial functions. Nonalgebraic functions often arise when trying to represent real-life phenomena mathematically. **Exponential** and **logarithmic functions** are important types of nonalgebraic functions.

Key 47 Exponential functions

OVERVIEW *The function* f(x) = x³ *is an example of a polynomial function. If the roles of the variable base x and the constant power 3 are interchanged, then the function that results,* F(x) = 3 x, *is called an **exponential** function.*

Exponential function: An exponential function has the form $f(x) = b^x$ where b is a positive constant different than 1. The domain of x is the set of real numbers. Since $f(x)$ has a positive value for all possible replacements of x, the range of an exponential function is limited to the set of positive real numbers.

Graphs of exponential functions: In Figure 8.1, we can compare the graphs of $y = b^x$ when $b > 1$ and when $0 < b < 1$. From the graphs, we observe the following:

- For $b > 1$, the function $y = b^x$ is an *increasing function,* rising as x increases. For $0 < b < 1$, $y = b^x$ is a *decreasing function*, falling as x increases.
- For $b > 1$ and for $0 < b < 1$, the graphs intersect the y-axis at (0, 1) but *do not* intersect the x-axis.
- The x-axis is an asymptote. For $b > 1$, the curve approaches the negative x-axis as x gets smaller. For $0 < b < 1$, the curve approaches the positive x-axis as x gets larger.

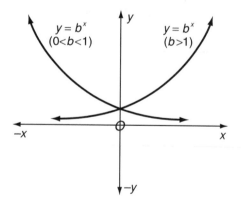

Figure 8.1 Graphs of $y = b^x$

Graphs of $y = b^x$ and $y = b^{-x}$ ($b > 0$ and $b \neq 1$): The graphs of $y = b^x$ and $y = b^{-x} = \left(\dfrac{1}{b}\right)^x$ are reflections of each other in the y-axis.

Figure 8.2 shows the graphs of $y = 2^x$ and $y = 2^{-x} = \left(\dfrac{1}{2}\right)^x$.

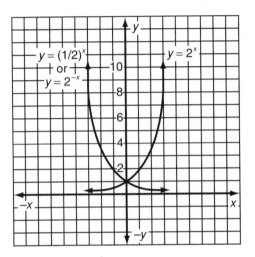

Figure 8.2 Graphs of $y = 2^x$ and $y = 2^{-x}$

One-to-one property of exponential functions: The graph of an exponential function $f(x) = b^x$ passes the horizontal line test so it is a one-to-one function. Thus, if $b^{x_1} = b^{x_2}$ then $x_1 = x_2$.

Exponential equation: An exponential equation is one in which the variable appears as part of an exponent. Some exponential equations can be solved by expressing each side as a power of the same base and then equating the exponents.

Example: If
$$8^{2x} = 16^{x+3},$$
then substituting 2^3 for 8 and 2^4 for 16 gives
$$2^{6x} = 2^{4x+12}.$$

By the one-to-one property of exponential functions, the exponents must be equal so
$$6x = 4x + 12$$
and
$$x = 6.$$

The natural base e: This special irrational number is denoted by the letter e in honor of the eighteenth century mathematician Leonhard Euler. The number e is approximately equal to 2.718 and can be estimated to any desired level of accuracy using the expression $\left(1 + \dfrac{1}{n}\right)^n$ for which larger and larger values of n give increasingly accurate approximations of e.

Natural exponential function: An exponential function whose base is e is a natural exponential function. These arise when certain types of growth processes are represented mathematically.

Key 48 Logarithmic functions

OVERVIEW *The inverse of the exponential function* $y = b^x$ *is the function* $x = b^y$. *In order to solve for* y *in terms of* x, *the* ***logarithmic function*** *was invented. If* $x = b^y$, *then* $y = \log_b x$ *which is read as* ***"logarithm of* x *to base* b *is* y."***

Definition of logarithmic function: The **logarithm** of a *positive* number x to a given base b, written as $\log_b x$, is the power to which b must be raised to obtain x, provided that $b > 0$ and $b \neq 1$. A logarithm is an exponent since

$$y = \log_b x \ \ means \ \ b^y = x$$

Graphs of logarithmic functions: Figures 8.3 and 8.4 compare the graphs of $y = b^x$ and its inverse function, $y = \log_b x$, for different values of their base. Notice that either graph is the reflection of the other in the line $y = x$. Table 8.1 compares the features of these graphs.

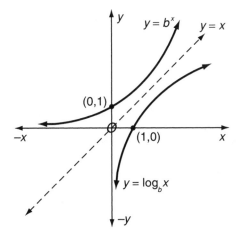

Figure 8.3 Graph of $y = b^x$ and $y = \log_b (b > 1)$

Table 8.1 Comparing $y = b^x$ and $y = \log_b x$

Feature	Graph of $y = b^x$	Graph of $y = \log_b x$
Quadrants of graph	I and II	I and IV
Domain	all real numbers	positive real numbers
Range	positive real numbers	all real numbers
Intercepts	(0, 1)	(1, 0)
Asymptote	x-axis	y-axis
$b > 1$	rises	rises
$0 < b < 1$	falls	falls

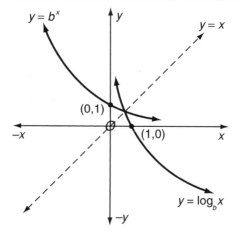

Figure 8.4 Graph of $y = b^x$ and $y = \log_b x$ $(0 < b < 1)$

Equivalent exponential and logarithmic forms: To convert between the two forms we use the fact that $y = \log_b x$ and $x = b^y$ mean the same thing. Here are some examples:

Exponential Form	Logarithmic Form
$4^3 = 64$	$\log_4 64 = 3$
$\sqrt{25} = 5$	$\log_{25} 5 = \dfrac{1}{2}$
$7^0 = 1$	$\log_7 1 = 0$
$3^{-2} = \dfrac{1}{9}$	$\log_3 \dfrac{1}{9} = -2$

Common logarithms: A common logarithm is a logarithm whose base is 10. A common logarithm is indicated when the base is omitted. Thus, $\log x$ means $\log_{10} x$. The common logarithm of a power of 10 is simply the exponent of 10. Thus, $\log 10 = 1$ since $10 = 10^1$; $\log 100 = 2$ since $100 = 10^2$; $\log 1000 = 3$ since $1000 = 10^3$; and so forth.

Natural logarithm: A natural logarithm, denoted by $\ln x$, is a logarithm whose base is e. Thus, $\ln x = \log_e x$. Here are some additional properties of the natural log function:

- $\ln e = 1$ since, in exponential form, $e^1 = e$.
- $\ln (e^x) = e^{\ln x} = x$ since the natural log and natural exponential functions are inverse functions.

KEY EXAMPLE

Solve for x: (a) $\log_x 8 = \dfrac{3}{2}$ (b) $\log_9 (27x) = x + 1$

Solution: Change each equation into exponential form and then solve the resulting equation.

(a) If

$$\log_x 8 = \frac{3}{2},$$

then

$$x^{3/2} = 8$$

and

$$x = 8^{2/3} = 4.$$

(b) If $\log_9 (27^x) = x + 1$

then

$$9^{x+1} = 27^x$$
$$3^{2(x+1)} = 3^{3x}$$
$$2(x+1) = 3x$$
$$x = 2$$

KEY EXAMPLE

Solve for x: $\ln e^{\sqrt{x}} = 2$.

Solution: Since $\ln (e^x) = x$,

$$\ln e^{\sqrt{x}} = \sqrt{x} = 2.$$

Thus,

$$x = 4.$$

Key 49 Logarithmic laws

OVERVIEW *Since logarithms are exponents, the laws for finding the logarithm of a product, quotient, and power are consistent with the laws of exponents for these operations.*

Product law of logarithms: The logarithm of a product equals the sum of the logarithms of its factors. Thus,

$$\log_b (xy) = \log_b x + \log_b y$$

Example:

$$\log_4 21 = \log_4 (7 \cdot 3) = \log_4 7 + \log_4 3$$

Quotient law of logarithms: The logarithm of a quotient equals the logarithm of the numerator minus the logarithm of the denominator. Thus,

$$\log_b\left(\frac{x}{y}\right) = \log_b x - \log_b y$$

Example:

$$\log\left(\frac{pq}{100}\right) = \log(pq) - \log 100 = \log p + \log q - 2$$

Power law of logarithms: The logarithm of a power of a quantity equals the product of the exponent and the logarithm of the quantity. Thus,

$$\log_b x^n = n \log_b x$$

Example 1:

$$\log \sqrt{10x} = \log(10x)^{1/2} = \frac{1}{2}\log 10x$$

$$= \frac{1}{2}(\log 10 + \log x)$$

$$= \frac{1}{2}(1 + \log x)$$

Example 2:

$$\ln e^2 + \ln \sqrt[3]{e} = 2\ln e + \frac{1}{3}\ln e$$

$$= 2(1) + \frac{1}{3}(1)$$

$$= \frac{7}{3}$$

KEY EXAMPLE

If $x = \log m$ and $y = \log n$, express $\log \sqrt[3]{\dfrac{m}{n}}$ in terms of x and y.

Solution:

$$\log \sqrt[3]{\frac{m}{n}} = \frac{1}{3}(\log m - \log n) = \frac{1}{3}(x - y)$$

KEY EXAMPLE

If $\log 2 = x$ and $\log 3 = y$, express each of the following in terms of x and y: (a) $\log 1.5$ (b) $\log \sqrt[3]{18}$

Solutions: (a) $\log 1.5 = \log \dfrac{3}{2} = \log 3 - \log 2 = y - x$.

(b) $\log \sqrt[3]{18} = \log(2 \cdot 3^2)^{\frac{1}{3}} = \dfrac{1}{3}(\log 2 + 2\log 3) = \dfrac{1}{3}(x + 2y)$

KEY EXAMPLE

If $\log N = 2 \log x + \log y$, express N in terms of x and y.

Solution: Work "backwards" in applying the logarithm laws for powers and products:

$$\log N = 2 \log x + \log y = \log x^2 + \log y = \log x^2 y$$

Since the logarithm function is one-to-one, $\log A = \log B$ implies $A = B$. Hence, $\log N = \log x^2 y$ means $N = x^2 y$.

Formula for a change of logarithm base: To change $\log_a c$ to an equivalent expression whose logarithm base is b, use the change of base formula

$$\log_a c = \frac{\log_b c}{\log_b a}.$$

For example, to express $\ln 7$ to a base 10 logarithm, let $a = e$, $c = 7$, and $b = 10$. Then

$$\ln 7 = \frac{\log_{10} 7}{\log_{10} e} = \frac{\log 7}{\log e}.$$

Key 50 Exponential and logarithmic
equations

OVERVIEW *Logarithm laws and properties can be used to solve exponential equations and equations that contain more than one logarithmic expression.*

Solving exponential equations using logarithms: If it is not possible to express both sides of an exponential equation as the power of the same rational base, then take the logarithm of each side of the equation.

Example: If $2^x = 6$, then

$$\log (2^x) = \log 6$$

Use the power law of logarithms:

$$x \log 2 = \log 6$$

Solve for x:

$$x = \frac{\log 6}{\log 2}$$

To find an *approximation* for x, evaluate the logarithms using a scientific calculator. Thus,

$$x = 0.7782 \div 0.3010 = 2.585$$

Solving logarithmic equations: Equations that contain logarithms are solved by eliminating the logarithms, either by writing the equation in exponential form or by consolidating the logarithms on both sides of the equation so the new equation has the form $\log_b N = \log_b M$ which implies $N = M$.

KEY EXAMPLE

Solve for x: $\log_3 (7x + 4) - \log_3 2 = 2 \log_3 x$

Solution: Use the reverse of the quotient law of logarithms to express the left side of the given equation as the logarithm of a single term. On the right side of the original equation use the reverse of the power law of logarithms. Then

$$\log_3 \left(\frac{7x + 4}{2} \right) = \log_3 x^2.$$

This equation has the form $\log_b N = \log_b M$, which implies $N = M$ where

$$N = \frac{7x + 4}{2} \quad \text{and} \quad M = x^2.$$

Thus,

$$\frac{7x + 4}{2} = x^2.$$

Multiplying each side of the equation by 2 and writing the resulting quadratic equation in standard form gives

$$2x^2 - 7x - 4 = 0$$

whose roots are $x = -\dfrac{1}{2}$ and $x = 4$. For $x = -\dfrac{1}{2}$ the right side of the original equation evaluates to the logarithm of a negative quantity which is not defined. Hence, the only root of the original equation is $x = 4$.

KEY EXAMPLE

Solve for x: $\log_4(x - 3) + \log_4(x + 3) = 2$

Solution: If

$$\log_4(x - 3) + \log_4(x + 3) = 2,$$

then

$$\log_4 [(x - 3)(x + 3)] = 2$$

and, in exponential form,

$$(x - 3)(x + 3) = 4^2.$$

Simplifying this equation gives $x^2 = 25$ so $x = -5$ or $x = 5$. Reject $x = -5$ since substituting this value for x in the original equation produces the logarithm of a negative quantity which is not defined. Thus, $x = 5$ is the only solution of the original equation.

KEY EXAMPLE

Solve for y in terms of e: $\ln 3y - \ln (y - 4) = 2$

Solution: Applying the reverse of the quotient law of logarithms on the left side of the given equation gives

$$\ln\left(\frac{3y}{y - 4}\right) = 2.$$

In exponential form,

$$\frac{3y}{y-4} = e^2.$$

Multiply each side of the equation by $y - 4$ to obtain

$$3y = e^2 y - 4e^2.$$

Collecting like terms and solving for y gives the solution

$$y = \frac{4e^2}{e^2 - 3}.$$

Key 51 Exponential growth and decay

OVERVIEW *The growth patterns of many types of biological and physical quantities closely approximate an exponential curve.*

Compound interest: If an amount of money A_0 is invested at a fixed annual rate r and interest is compounded n times each year, then the amount of money in t years, denoted by $A(t)$, is given by the equation

$$A(t) = A_0 \left(1 + \frac{r}{n}\right)^{nt}$$

Example:

If \$1,200 is invested at an annual rate of 8% for 5 years and interest is compounded quarterly each year, then $A_0 = 1200$, $r = 0.08$, $t = 5$, and $n = 4$. Thus,

$$A(5) = 1200(1 + 0.02)^{5(4)} = 1200(1.02)^{20} = 1783.14$$

In 5 years the investment grows from \$1,200 to \$1,783.14.

A model of natural exponential growth: An equation of the form

$$A(t) = A_o \, e^{kt}$$

predicts the growth of a subtance where A_0 is the amount of the substance present at $t = 0$ and $A(t)$ is the amount that will be present at time t. The number k is a constant whose value depends on the specific nature of the growth process. For example, if an amount of money A_0 is invested at a fixed annual rate r and interest is *continuously* compounded for t years, then $A(t) = A_0 \, e^{rt}$.

Example:

If \$1,200 is invested at an annual rate of 8% for 5 years and interest is compounded continuously, then

$$A(5) = 1200 \, e^{(0.08)5} = 1200 e^{0.40} = 1790.19$$

Thus, in 5 years the investment grows from \$1,200 to \$1,790.19.

Model of natural exponential decay: If k is negative, an equation of the form $A(t) = A_0 \, e^{kt}$, models exponential decay since $A(t)$ *decreases* over time. The disintegration of radioactive elements follows this type of law.

KEY EXAMPLE

If \$2,750 is invested at an annual rate of 9%, find the number of years it will take for the investment to double if interest is continuously compounded.

Solution: Use the formula $A(t) = A_0 \, e^{rt}$. where $A_0 = 2750$, $r = 0.09$, and $A(t) = 2 (2750) = 5500$. Thus,

$$5500 = 2750 \, e^{0.09t}.$$

Dividing both sides of the equation by 2750 gives

$$2 = e^{0.09t}.$$

In logarithmic form the equation is

$$0.09t = \ln 2.$$

Thus,

$$t = \frac{\ln 2}{0.09}.$$

Using a calculator, $\ln 2$ is approximately 0.693 so

$$t = 0.693 \div 0.09 = 7.7.$$

The investment will double in approximately 7.7 years.

KEY EXAMPLE

The number of bacteria in a certain culture is 900. If the bacteria grows in the culture according to the law $A(t) = A_0 \, e^{0.5t}$, what will be the number of bacteria in the culture 2 hours from now?

Solution: At $t = 0$, $A_0 = 900$. Let $t = 2$. Then

$$A(2) = 900 \, e^{0.5(2)} = 2446.45.$$

Two hours from now the number of bacteria in the culture will be approximately 2446.

KEY EXAMPLE

A radioactive element decays exponentially according to the law $A(t) = A_0 \, e^{kt}$. After 200 days a 25 mg sample of this element weighs 6 mg. Find the value of the decay constant k.

Solution: At $t = 0$, $A_0 = 25$. Let $t = 200$ and $A(200) = 6$. Then,

$$6 = 25 \, e^{k(200)}$$

so

$$e^{200k} = \frac{6}{25}.$$

In logarithmic form,

$$200k = \ln\left(\frac{6}{25}\right).$$

Thus,

$$k = \frac{\ln 6 - \ln 25}{200} = \frac{1.792 - 3.219}{200} = -0.007135$$

KEY EXAMPLE

The half-life of a radioactive substance is the amount of time it takes for one-half of the substance to disintegrate. If a radioactive substance decays according to the law $A(t) = A_0\, e^{kt}$, where t is measured in days, find its half-life to the nearest day if $k = -0.025$.

Solution: In the formula $A(t) = A_0\, e^{kt}$, let $A(t) = 0.5\, A_0$, $k = -0.025$, and then solve for t. Thus,

$$0.5\, A_0 = A_0\, e^{(-0.025)t}$$

or

$$0.5 = e^{-0.025t}.$$

In logarithmic form the equation is

$$-0.025t = \ln 0.5.$$

Using a calculator, we obtain -0.693 as an approximation for $\ln 0.5$. Thus,

$$t = -0.693 \div (-0.025) = 27.72.$$

The half-life to the nearest day is 28.

Theme 9 CONIC SECTIONS AND
THEIR EQUATIONS

A *circle, ellipse, parabola,* and *hyperbola* are plane curves, called **conic sections**, that can be formed by cutting a double cone with a plane at various angles, as shown in Figure 9.1. A **circle** is formed when the plane intersects one cone and is perpendicular to the axis; if the plane is not perpendicular to the axis, an **ellipse** is formed. A **parabola** is formed when the plane intersects one cone and is parallel to an edge of the cone. A **hyperbola** is formed when a plane intersects both cones.

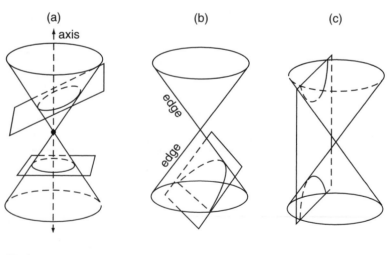

(a)	(b)	(c)
Circle and Ellipse	Parabola	Hyperbola

Figure 9.1 Conic Sections.

Conic sections arise in diverse areas of applied mathematics. They can be used to help describe the paths of projectiles, planetary motion, and the shapes of light reflecting surfaces. Each of the conic section curves can be defined geometrically and its equation derived using the methods of analytic (coordinate) geometry.

INDIVIDUAL KEYS IN THIS THEME

52	The circle
53	The ellipse
54	The hyperbola
55	The parabola
56	The general equation for conic sections

Key 52 The circle

OVERVIEW *A circle is defined geometrically in terms of a fixed point in a plane and a number.*

Circle: A circle is the set of all points P that are a fixed distance r, called the **radius**, from a given point (h, k), called the **center**. In the accompanying figure, $P(x, y)$ is on the circle if and only if its distance from (h, k) equals r:

$$\sqrt{(x - h)^2 + (y - k)^2} = r$$

Squaring both sides of this equation gives the standard form of an equation of a circle,

$$(x - h)^2 + (y - k)^2 = r^2.$$

Thus, the graph of the equation $x^2 + y^2 = r^2$ is a circle whose radius is r and whose center is at $(0, 0)$.

KEY EXAMPLE

Write an equation of a circle with the following:

(a) a radius length of 5 and center at $(-1, 4)$.
(b) a diameter whose endpoints are $A(-7, 5)$ and $B(1, 5)$.

Solution: (a) In the equation

$$(x - h)^2 + (y - k)^2 = r^2,$$

let $(h, k) = (-1, 4)$ and $r = 5$ so

$$[x - (-1)]^2 + (y - 4)^2 = 5^2.$$

Simplifying gives

$$(x + 1)^2 + (y - 4)^2 = 25$$

as an equation of a circle with radius 5 and center $(-1, 4)$.

(b) The midpoint of AB is $(-3, 5)$ which is the center of the circle. Using the distance formula you can verify $AB = 8$ so the radius is 4. The equation of a circle with a radius length of 4 and center at $(-3, 5)$ is

$$(x + 3)^2 + (y - 5)^2 = 16.$$

Key 53 The ellipse

OVERVIEW *Given two fixed points* F_1 *and* F_2 *and a positive number* k, *an **ellipse** is traced out by connecting the set of all points* P *in the plane that satisfy the condition that* $PF_1 + PF_2 = k$ *(see Figure 9.2).*

Definition of ellipse: An ellipse is the set of all points in the plane such that the sum of the distances from two fixed points, F_1 and F_2, is constant. The two fixed points are called the **foci** (plural of focus).

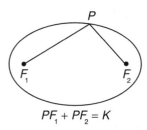

Figure 9.2 Ellipse Definition   **Figure 9.3** Ellipse Terms

Terms associated with an ellipse:

- **Center:** the midpoint of the segment whose endpoints are the two foci. The distance from the center to either focus is represented by c.
- **Vertices:** the points V_1 and V_2 on the ellipse that are the endpoints of the line segment that contains the foci. The distance from the center to either of the two vertices is denoted by a.
- **Axes of symmetry:** an ellipse has two axes of symmetry that pass through its center. The **major axis** is the line segment whose endpoints are V_1 and V_2. The **minor axis** is the line segment that is perpendicular to the major axis and whose endpoints are points on the ellipse each of which is b units from the center where $b^2 = a^2 - c^2$ $(a > b > 0)$. Since the length of the major axis is $2a$ and the length of the minor axis is $2b$, the major axis is always the longer of the two axes of symmetry. If the axes of symmetry have the same length, then the ellipse is a circle.

Ellipse equations with center at (0, 0): We can compare ellipses with horizontal and vertical major axes in Figures 9.4 and 9.5. Each ellipse is labeled with its equation. The foci are located at $F(\pm c, 0)$ for a horizontal ellipse and at $F(0, \pm c)$ for a vertical ellipse.

Figure 9.4 Horizontal Ellipse

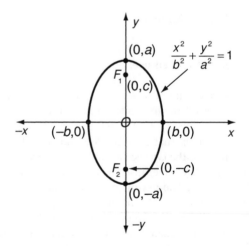

Figure 9.5 Vertical Ellipse

Ellipse equations with center at (h, k): The standard form of the equation of an ellipse whose center is at (h, k) is given in Table 9.1 where $b^2 = a^2 - c^2$ $(a > b > 0)$ or, equivalently, $c = \sqrt{a^2 - b^2}$.

Table 9.1 Ellipse Equations where $c = \sqrt{a^2 - b^2}$ **and** $a > b > 0$

Major Axis	Equation	Foci	Vertices
Horizontal	$\dfrac{(x-h)^2}{a^2} + \dfrac{(y-k)^2}{b^2} = 1$	$F(h \pm c, k)$	$V(h \pm a, k)$
Vertical	$\dfrac{(x-h)^2}{b^2} + \dfrac{(y-k)^2}{a^2} = 1$	$F(h, k \pm c)$	$V(h, k \pm a)$

KEY EXAMPLE

Find the coordinates of the foci and vertices of an ellipse whose equation is $\dfrac{x^2}{16} + \dfrac{y^2}{9} = 1$.

Solution: Since the x^2-term has the larger denominator, the major axis is horizontal. Let $a^2 = 16$ and $b^2 = 9$. Then $a = 4$, $b = 3$, and $c = \sqrt{a^2 - b^2} = \sqrt{7}$. Since the center is at the origin and the major axis coincides with the x-axis, the foci are at $F(\pm c, 0) = F_1(-\sqrt{7}, 0)$ and $F_2(\sqrt{7}, 0)$ and the vertices are at $V(\pm a, 0) = V_1(-4, 0)$ and $V_2(4, 0)$.

KEY EXAMPLE

Find the coordinates of the foci and vertices of an ellipse whose equation is $25x^2 + 4y^2 = 100$.

Solution: Put the given equation in standard form by dividing each term of the equation by 100 which gives

$$\frac{x^2}{4} + \frac{y^2}{25} = 1.$$

Since the y^2-term has the larger denominator and the center is at the origin, the major axis is vertical and lies along the y-axis. Let $a^2 = 25$ and $b^2 = 4$. Then $a = 5$, $b = 2$, and $c = \sqrt{a^2 - b^2} = \sqrt{21}$. Hence, the foci are at $F(0, \pm c) = F_1(0, -\sqrt{21})$ and $F_2(0, \sqrt{21})$ and the vertices are at $V(0, \pm a) = V_1(0, -5)$ and $V_2(0, 5)$.

KEY EXAMPLE

For the ellipse whose foci are (−1, 2) and (7, 2) and whose minor axis has a length of 6, find the center, the vertices, and the standard form of its equation.

Solution: Since the foci have the same *y*-coordinate, the major axis is horizontal. The center (*h, k*) of the ellipse is the midpoint of the segment whose endpoints are the foci. Hence,

$$(h, k) = \left(\frac{-1+7}{2}, \frac{2+2}{2} \right) = (3, 2).$$

Since the distance from either focus to (3, 2) is 4, *c* = 4. The length of the vertical minor axis is given as 6, so 2*b* = 6 and *b* = 3. Thus,

$$a = \sqrt{c^2 + b^2} = \sqrt{4^2 + 3^2} = 5.$$

The vertices are at *V*(*h* ± *a, k*) = *V*(3 ± 5, 2) = V_1(−2, 2) and V_2(8, 2). Letting (*h, k*) = (3, 2), *a* = 5, and *b* = 3 in the standard form of the equation of a horizontal ellipse gives

$$\frac{(x-3)^2}{25} + \frac{(y-2)^2}{9} = 1$$

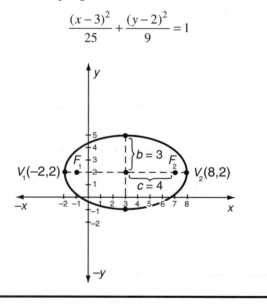

Key 54 The hyperbola

OVERVIEW *An ellipse is formed by connecting the set of all points* P *such that the* **sum** *of the distances from the foci to* P *are constant. If the* **difference** *of these distances is taken, a* **hyperbola** *is formed.*

Definition of hyperbola: A hyperbola is the set of all points *P* in the plane such that the difference of the distances from *P* to two fixed points, called **foci**, is constant. As shown in Figure 9.6, a hyperbola consists of two nonintersecting curves called **branches**. The branches resemble a pair of parabolas that open in opposite directions.

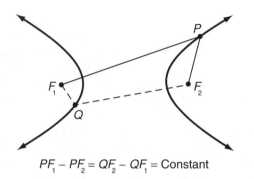

$$PF_1 - PF_2 = QF_2 - QF_1 = \text{Constant}$$

Figure 9.6   Hyperbola Definition

Terms associated with a hyperbola:

- **Center:** the midpoint of the segment whose endpoints are the foci. As with an ellipse, *c* represents the distance from the center to a focus point.
- **Vertices:** the points V_1 and V_2 at which the line that contains the foci intersects the branches of the hyperbola. The distance from the center to either of the two vertices is denoted by *a*. Unlike an ellipse, $c > a$ for a hyperbola.

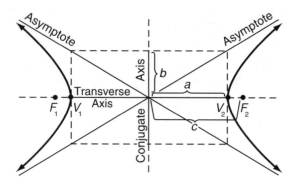

Figure 9.7 Hyperbola Terms

- **Axes of symmetry:** like an ellipse, a hyperbola has two axes of symmetry that pass through its center. The **transverse axis** is the line segment whose endpoints are V_1 and V_2. The **conjugate axis** is the line segment that is perpendicular to the transverse axis and whose endpoints are each b units from the center where a, b, and c are related by the equation $b^2 = c^2 - a^2$ ($c > a > 0$). The length of the transverse axis is $2a$, and the conjugate axis measures $2b$. Unlike an ellipse, either axis of symmetry may be longer than the other, or both axes may have the same length.
- **Hyperbola asymptotes:** two lines that the branches of the curve approach as they depart from the center. The asymptotes contain the diagonals of the rectangle with the same center as the hyperbola and whose adjacent sides measure $2a$ and $2b$.

Hyperbola equations with center at (0, 0): Figures 9.8 and 9.9 show hyperbolas with horizontal and vertical transverse axes. Each hyperbola is labeled with its equation. The foci are located at $F(\pm c, 0)$ for a horizontal hyperbola and at $F(0, \pm c)$ for a vertical hyperbola. The equations of the asymptotes are $y = \pm \dfrac{b}{a} x$ for a horizontal hyperbola, and $y = \pm \dfrac{a}{b} x$ for a vertical hyperbola.

Figure 9.8   Horizontal Hyperbola

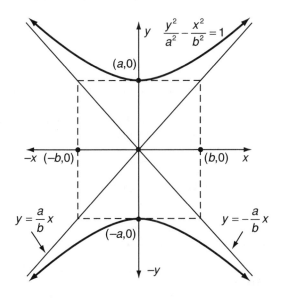

Figure 9.9   Vertical Hyperbola

Hyperbola equations with center at (h, k): The standard form of the equation of a hyperbola whose center is at (h, k) is given in Table 9.2 where $b^2 = c^2 - a^2$ $(c > a > 0)$ or, equivalently, $c = \sqrt{a^2 + b^2}$.

Table 9.2 Hyperbola Equations

	Horizontal Transverse Axis	Vertical Transverse Axis
Equation	$\dfrac{(x-h)^2}{a^2} - \dfrac{(y-k)^2}{b^2} = 1$	$\dfrac{(y-k)^2}{a^2} - \dfrac{(x-h)^2}{b^2} = 1$
Foci	$F(h \pm c, k)$	$F(h, k \pm c)$
Vertices	$V(h \pm a, k)$	$V(h, k \pm a)$
Asymptotes	$y - k = \pm \dfrac{b}{a}(x - h)$	$y - k = \pm \dfrac{a}{b}(x - h)$

KEY EXAMPLE

What is an equation of a hyperbola whose vertices are at $(\pm 12, 0)$ and whose foci are at $(\pm 13, 0)$?

Solution: Since the vertices lie on the x-axis and are the endpoints of a segment whose midpoint is the origin, the hyperbola has a horizontal transverse axis and its center at $(h, k) = (0, 0)$. The vertices of this type of hyperbola are at $V(\pm a, 0) = V(\pm 12, 0)$ and the foci are at $F(\pm c, 0) = F(\pm 13, 0)$. Since $a = 12$ and $c = 13$,

$$b = \sqrt{c^2 - a^2} = \sqrt{13^2 - 12^2} = 5.$$

Thus, the standard equation of this horizontal hyperbola is

$$\frac{x^2}{144} - \frac{y^2}{25} = -1.$$

KEY EXAMPLE

What are the coordinates of the center, vertices, and foci of the hyperbola whose equation is $\dfrac{(y-1)^2}{9} - \dfrac{(x+4)^2}{7} = 1$?

Solution: The equation has the form

$$\frac{(y-k)^2}{a^2} - \frac{(x-h)^2}{b^2} = 1$$

where $h = -4$, $k = 1$, $a^2 = 9$, $b^2 = 7$, and $c = \sqrt{a^2 + b^2} = \sqrt{25} = 5$. Thus, the hyperbola has a vertical transverse axis with center = $(h,\ k)$ = $(-4, 1)$, vertices = $V(h,\ k \pm a) = V(-4, 1 \pm 3) = V_1(-4, -2)$ and $V_2(-4, 4)$, and foci = $F(h,\ k \pm c) = F(-4, 1 \pm 5) = F_1(-4, -4)$ and $F_2(-4, 6)$.

KEY EXAMPLE

Find the standard equation of the hyperbola whose foci are at $(-1, -3)$ and $(-1, 7)$ and whose conjugate axis has a length of 8. Sketch its graph.

Solution: First determine the equation of the hyperbola. The foci are on the same vertical line since they have the same x-coordinate. Hence, the hyperbola has a transverse axis whose length = $7 - (-3) = 10$. The center $(h,\ k)$ is the midpoint of the segment whose endpoints are the foci. Hence

$$(h, k) = \left(\frac{-1 + (-1)}{2}, \frac{-3 + 7}{2} \right) = (-1, 2)$$

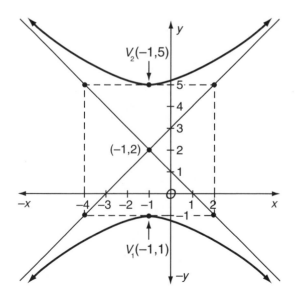

The distance c from the center to either of the two foci is 5. Since the length of the conjugate axis is 8, $2b = 8$, so $b = 4$. Thus,

$$a = \sqrt{c^2 - b^2} = \sqrt{5^2 - 4^2} = 3.$$

Since h = –1, $k = 2$, $a^2 = 9$, and $b^2 = 16$, the standard equation of this vertical hyperbola is

$$\frac{(y-2)^2}{9} - \frac{(x+1)^2}{16} = 1.$$

Graph the center $(h, k) = (-1, 2)$ and the vertices $V(h, k \pm a) = V(-1, 2 \pm 3) = V_1(-1, -1)$ and $V_2(-1, 5)$. Draw a rectangle with base $2b = 8$ and height $2a = 6$ such that the vertices are the midpoints of a pair of opposite sides of the rectangle. Then draw the asymptotes through the diagonals of the rectangle. Sketch the hyperbola, using the asymptotes as guideposts, as the branches of the curve move away from the vertices.

Rectangular hyperbolas: The graph of the equation $xy = k$ $(k \neq 0)$ [see Key 33, page 91] is a special type of hyperbola, called a **rectangular** or **equilateral hyperbola**. The foci of this type of hyperbola lie on the line $y = x$.

Key 55 The parabola

OVERVIEW *Given a line and a fixed point not on the line, a **parabola** is traced out by connecting the set of all points* Q *whose distance from the line and from the fixed point is the same.*

Parabola: A parabola is the set of all points Q in the plane such that the distance from Q to a fixed point F, called the **focus**, is equal to the distance from Q to a given line ℓ, called the **directrix**.

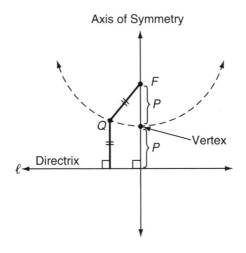

Axis of symmetry: The axis of symmetry is line the that contains the focus and is perpendicular to the directrix.

Vertex (turning point): The point V at which the axis of symmetry intersects the parabola is called the vertex. The directrix and the focus are on opposite sides of the vertex and are the same distance from the vertex. The directed distance from the vertex to either the focus or directrix is denoted by p.

Parabola equations: The standard form of the equation of a parabola whose vertex is (h, k) is given in Table 9.3.

Table 9.3 The Standard Form of Parabola Equations

Parabola Equation	$p > 0$	$p < 0$
$(x - h)^2 = 4p(y - k)$ where • axis of symmetry: $x = h$ • focus: $F(h, k + p)$ • directrix: $y = k - p$		
$(y - k)^2 = 4p(x - h)$ where • axis of symmetry: $y = k$ • focus: $F(h + p, k)$ • directrix: $x = h - p$		

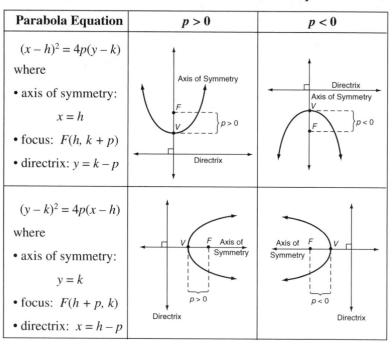

KEY EXAMPLE

Write an equation of the parabola whose vertex is at $V(0, 0)$ and whose focus is at $F(0, 3)$. What is an equation for the axis of symmetry and for the directrix?

Solution: The vertex and focus lie on the same vertical line (the *y*-axis) so the standard equation of the parabola has the form

$$(x - h)^2 = 4p(y - k)$$

where $(h, k) = (0, 0)$. Since the focus is located 3 units *above* the vertex, $p = 3$. Hence, an equation of this vertical parabola is

$$x^2 = 12y.$$

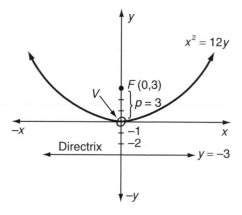

An equation of the axis of symmetry is

$$x = h = 0.$$

The directrix is $|p|$ units on the opposite side of the vertex and is perpendicular to the axis of symmetry. Since $p = 3$, an equation of the directrix is

$$y = k - p = 0 - 3 = -3.$$

KEY EXAMPLE

Determine the coordinates of the vertex and the focus of the parabola whose equation is $y^2 - 6y + 8x + 17 = 0$. Give equations for the axis of symmetry and for the directrix.

Solution: The given equation can be put into the form

$$(y - k)^2 = 4p(x - h)$$

by using the method of completing the square. Isolating terms involving y in the original equation gives

$$y^2 - 6y = -8x - 17.$$

Complete the square by adding the square of one-half of -6 to both sides of the equation. Thus,

$$y^2 - 6y + 9 = -8x - 17 + 9$$
$$(y - 3)^2 = -8x - 8$$
$$(y - 3)^2 = -8(x + 1)$$

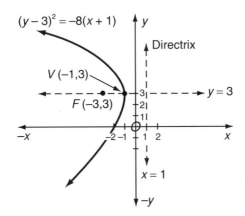

The equation $(y - 3)^2 = -8(x + 1)$ has the form $(y - k)^2 = 4p(x - h)$ where $h = -1$, $k = 3$, and $4p = -8$ so $p = -2$. Thus, the vertex $V(h, k) = (-1, 3)$. Since p is negative, the parabola opens to the left and the focus is 2 units to left of $(-1, 3)$. Hence, the focus is at

$$F(h + p, k) = F(-1 + (-2), 3) = F(-3, 3).$$

An equation of the horizontal axis of symmetry is

$$y = k = 3.$$

An equation of the vertical directrix is

$$x = h - p = -1 - (-2) = 1.$$

Key 56   The general equation

for conic sections

OVERVIEW   *Different conic sections can be described by the same general equation.*

Conic section equations with center at (0, 0): The curve traced out by the equation $Ax^2 + By^2 = C$ $(C \neq 0)$ depends on the values of the constants A, B, and C.

1. The graph is a **circle**, if A and B are equal and have the same sign as C;
2. The graph is an **ellipse**, if A and B are unequal and have the same sign as C;
3. The graph is a **hyperbola**, if A and B have different signs.

Conic section equations with center at (h, k): If the standard equations for a circle, ellipse, hyperbola, and parabola are expanded and simplified, then each of the resulting equations will have the form $Ax^2 + By^2 + Cx + Dy + E = 0$ where A, B, C, D, and E are real numbers with at least one of coefficients A and B not 0.

Degenerate conics: For some special sets of values of the coefficients of the equation $Ax^2 + By^2 + Cx + Dy + E = 0$, the graph of the equation is a **degenerate conic**—a point, line, or pair of lines. For example, the graph of $x^2 + y^2 = 0$ is the point (0, 0), and the graph of $x^2 - y^2 = 0$ is a pair of intersecting lines whose equations are $y = x$ and $y = -x$. For some numerical coefficients, the general equation may not have any real solution pairs so the graph is the empty set, as in $x^2 + y^2 + 16 = 0$.

Identifying a conic section from its equation: If the graph of $Ax^2 + By^2 + Cx + Dy + E = 0$ is not a degenerate conic, then its graph can be identified by comparing A and B:

- if exactly one of A or B is 0, the graph is a parabola.
- if $A = B$, the graph is a circle.
- if $A \neq B$ and A and B have the same sign, the graph is an ellipse.
- if A and B have opposite signs, the graph is a hyperbola.

KEY EXAMPLE

Name the conic section whose equation is $x^2 - 9y^2 - 10x - 54y - 47 = 0$ and determine the coordinates of its center, vertices, and foci.

Solution: The given equation has the form $Ax^2 + By^2 + Dx + Ey + F = 0$ where A and B have opposite signs. Hence, its graph is a hyperbola.

To find its center, put the equation into standard form by completing the square for x and for y. Grouping like variable terms together gives

$$(x^2 - 10x + ?) - 9(y^2 + 6y + ?) = 47$$

Completing the square for each quadratic polynomial yields

$$(x^2 - 10x + 25) - 9(y^2 + 6y + 9) = 47 + 25 - 81$$
$$(x - 5)^2 - 9(y + 3)^2 = -9$$

To put this equation in standard form, divide each member of the equation by –9. Thus,

$$\frac{(y + 3)^2}{1} - \frac{(x - 5)^2}{9} = 1.$$

The derived hyperbola equation has the form

$$\frac{(y - k)^2}{a^2} - \frac{(x - h)^2}{b^2} = 1$$

in which $h = 5$, $k = -3$, $a^2 = 1$, $b^2 = 9$, and $c = \sqrt{a^2 + b^2} = \sqrt{1 + 9} = \sqrt{10}$. The graph is a vertical hyperbola with center at $(h,\ k) = (5, -3)$. The vertices are at $(h,\ k \pm a) = (5, -3 \pm 1) = V_1(5, -2)$ and $V_2(5, -4)$, and the foci are at $(h,\ k \pm c) = F_1(5, -3 - \sqrt{10})$ and $F_2(5, -3 + \sqrt{10})$.

Theme 10 COUNTING METHODS

*A*n important application of mathematics is counting the number of ways in which the members of a set can be arranged in a certain order or can be selected as members of a subset. Given enough time, these arrangements or selections can be counted by listing the members for each possible arrangement or selection. More efficient counting methods make use of formulas for **permutations** (ordered arrangements of objects) and **combinations** (unordered subsets of objects).

Key 57 The basic principles of counting

OVERVIEW *If we know the number of ways in which each of two events can occur, we can find the number of ways in which both events or one of the two events can happen.*

Fundamental rules of counting: Assume event A can occur in a ways and event B can occur in b ways.

- **Multiplication principle of counting:** events A and B can both occur in $a \times b$ ways. This rule can be extended to more than two events.

  Example 1: If Bill has 6 different shirts, 4 different ties, and 3 different sport jackets, then the number of different outfits he can choose, which consist of a shirt, tie, and sport jacket is $6 \cdot 4 \cdot 3 = 72$.

  Example 2: The total number of possible arrangements of the three letters A, B, and C can be determined in the following way: The first position can be filled by any of the *three* letters. Any of the remaining *two* letters can occupy the middle position. Once the first and second positions are determined, there is *one* letter left that can be put in the last position. By the multiplication principle of counting, the total number of arrangements is $3 \cdot 2 \cdot 1 = 6$. Similarly, four letters can be arranged in $4 \cdot 3 \cdot 2 \cdot 1 = 24$ ways, five letters can be arranged in $5 \cdot 4 \cdot 3 \cdot 2 \cdot 1 = 120$ ways, and so forth.

- **Addition principle of counting:** events A or B can occur in $a + b$ ways, provided the two events are **mutually exclusive**. This means that the two events cannot occur at the same time, just as we cannot obtain a head *and* a tail when a coin is flipped.

Key 58 Permutations

OVERVIEW *The number of different ways in which objects can be arranged in a row can be counted using a simple formula.*

Permutation: A permutation is an arrangement of objects in a definite order.

Example: The set of letters {*A, B, C*} can be arranged in six different orders:

$$ABC, ACB, BAC, BCA, CAB, \text{ and } CBA.$$

Each three-letter arrangement represents a *permutation* of the three letters.

n **Factorial:** The product of consecutive positive integers from *n* down to 1 is called *n* factorial and is denoted by *n*!. By definition, 0! = 1.

Example: $5! = 5 \cdot 4 \cdot 3 \cdot 2 \cdot 1 = 120$.

$_nP_r$: Denotes the number of different ordered arrangements of *r* out of *n* objects where $0 \leq r \leq n$. The symbol $_nP_r$ is read as, "the number of permutations of *n* things taken *r* at a time." Thus, $_nP_n = n!$ is the number of permutations when all *n* objects are used in each arrangement.

Permutation formula: If *r* and *n* are positive integers with $r \leq n$, then

$$_nP_r = \frac{n!}{(n-r)!} = \underbrace{n(n-1)(n-2)...(n-r+1)}_{\substack{product \ of \ the \\ r \ greatest \ factors \ of \ n}}$$

If $r = n$, the formula simplifies to

$$_nP_n = n!$$

Example: The number of different three-digit numbers that can be formed using digits 1, 2, 3, 4, 5, and 6, assuming no digit is used more than once, is

$$_6P_3 = 6 \cdot 5 \cdot 4 = 120.$$

If the digits can be used more than once, then the number of different three-digit numbers that can be formed is, by the multiplication principle of counting, $6 \times 6 \times 6 = 216$.

Permutations with conditions attached: Sometimes arrangements are formed subject to special conditions. For example, suppose we want to know the number of three-digit numbers greater than 500 that can be formed using the digits 1, 2, 3, 4, 5, 6, and 7. The digit used for the first position of each arrangement must satisfy the condition that it is 5 or greater. There are three such digits (5, 6, and 7). Assuming no repetition of digits, the second position of each 3-digit arrangement may be filled by any of the remaining six digits and the last position may be filled by any of the remaining five digits. By the multiplication principle of counting, the total number of three-digit numbers greater than 500 that can be formed is $3 \times 6 \times 5 = 90$. We could have also reasoned that the first position may be filled in $_3P_1$ ways and the remaining two positions filled in $_6P_2$ ways. The number of three-digit numbers greater than 500 is $_3P_1 \cdot {_6P_2} = 90$. If repetition of digits is allowed, then $3 \times 7 \times 7 = 147$ three-digit numbers greater than 500 can be formed.

Addition principle and permutations: Permutations may need to be added to obtain the total number of possible arrangements.

Example: How many numbers greater than 400 and less than 9999 can be formed using the digits 0, 1, 3, 4, and 5 if no digit is used more than once? Since the leading digit must be either a 4 or a 5, there are $2 \cdot 4 \cdot 3 = 24$ three-digit numbers greater than 400 that can be formed. Every four-digit number is greater than 400. However, since 0 cannot be used for the first digit, $4 \cdot 4 \cdot 3 \cdot 2 = 96$ four-digit numbers can be formed. Thus, the number of three digit or four digit numbers greater than 400 that can be formed is $24 + 48 = 120$.

Circular permutations: The number of ways in which n different objects can be arranged in a circle is $(n - 1)!$

Example: Five people can be seated at a round table in $(5 - 1)! = 4! = 24$ different ways.

Permutations of objects with some identical: If in a set of n objects, a objects are identical, b objects are identical, c objects are identical, and so forth, then the number of different ways in which the n objects taken all at a time can be arranged is

$$\frac{n!}{a!\,b!\,c!\,\ldots}$$

Example: The number of different ways four red flags, three blue flags, and one green flag can be arranged on a vertical flagpole is

$$\frac{(4+3+1)!}{4!\,3!\,1!} = \frac{8!}{4!\,3!} = \frac{8 \cdot 7 \cdot \overset{1}{\cancel{6}} \cdot 5 \cdot \overset{1}{\cancel{4!}}}{\cancel{4!} \cdot 3 \cdot 2 \cdot 1} = 280.$$

Key 59 Combinations

OVERVIEW *Although there are six **permutations** of the three letters* A, B, *and* C, *there is only one distinct set or **combination** of the three letters*—{A, B, C}.

$_nC_r$: This symbol denotes the number of subsets with r members that can be formed from a set with n members where $0 \le r \le n$. The symbol $_nC_r$ is read as, "the number of combinations of n things taken r at a time."

Combination formula: If $0 \le r \le n$, then

$$_nC_r = \frac{_nP_r}{r!} = \frac{n!}{r!(n-r)!}$$

Combinations vs. permutations: Permutations are *ordered* arrangements of objects while **combinations** are collections of objects in which only the identity of the objects matters. For example, suppose three people are chosen from a group of five people. If the people stand on a line, then their order is significant and the number of different ordered arrangements of 3 out of 5 people is $_5P_3 = 5 \cdot 4 \cdot 3 = 60$. If the three people are selected to serve on a committee, then their names rather than their order is important, so the total number of three-member committees that can be formed is $_5C_3$ where

$$_5C_3 = \frac{5!}{3!(5-2)!} = \frac{5!}{3!2!} = \frac{5 \cdot \overset{2}{\cancel{4}} \cdot \overset{1}{\cancel{3!}}}{\cancel{3!} \cdot \cancel{2} \cdot 1} = 10 .$$

Combinations for event *A AND* **event** *B*: Multiplying the number of combinations for event *A* by the number of combinations for event *B* gives the number of selections for the two events occuring together.

Example: There are six pens and seven books on a desk. If a student wants to take four pens (event *A*) *and* three books from the desk (event *B*), how many selections are possible? Four pens can be selected from six pens in $_6C_4$ ways. Three books can be selected from seven books in $_7C_3$ ways. Thus, the number of selections that include four pens *and* three books is $_6C_4 \cdot _7C_3 = 525$.

Combinations for event *A OR* **event** *B*: Adding the number of combinations for event *A* to the number of combinations for a mutually exclusive event *B* gives the number of selections for exactly one of the two events occurring.

Example: A jar contains four blue marbles and six green marbles. If three marbles are drawn without looking, in how many ways can the three marbles be selected so that they have the same color? The three marbles will have the same color if three blue marbles are selected (event *A*) *or* if three green marbles are selected (event *B*). Three blue marbles can be selected from four blue marbles in $_4C_3$ ways. Three green marbles can be selected from six green marbles in $_6C_3$ ways. Thus, the total number of ways in which three marbles having the same color can be selected is

$$_4C_3 + _6C_3 = 4 + 20 = 24.$$

Combinatorial relationships: These formulas can often save time in evaluating combinations that have the indicated form:

- $_nC_n = 1$
- $_nC_1 = n$
- $_nC_0 = 1$
- $_nC_k = _nC_{n-k} \quad (k \le n)$

Combinations taken from 1 to *n*: The sum of the combinations of *n* different objects taken 1, 2, 3, . . . , and *n* at a time is $2^n - 1$. For example, the total number of different subcommittees consisting of at least one person that can be selected from a group of 5 people is $2^5 - 1 = 31$.

KEY EXAMPLE

From a committee of seven Republicans and five Democrats a five-member subcommittee is to be selected. In how many different ways can the subcommitte be selected if *at least* three of its members must be Democrats?

Solution: The subcommitte will have *at least* three Democrats if it includes three, four, or five Democrats.

A committee of 3 Democrats *and* 2 Republicans can be selected $_5C_3 \cdot _7C_2 = 210$ ways.
A committee of 4 Democrats *and* 1 Republican can be selected $_5C_4 \cdot _7C_1 = 35$ ways.
A committee of 5 Democrats *and* 0 Republicans can be selected $_5C_5 \cdot _7C_0 = 1$ way.

Thus, there are is a total of 210 + 35 + 1 ways in which a five-member subcommittee that includes at least three Democrats can be selected.

Key 60   The Binomial Theorem

OVERVIEW   *Expanding a binomial to a positive integer power, as in*

$$(a + b)^5 = a^5 + 5a^4b + 10a^3b^2 + 10a^2b^3 + 5ab^4 + b^5$$

*can be accomplished using a formula called the **binomial theorem**.*

Binomial theorem: For any positive integer n, the expansion of $(a + b)^n$ is given by the formula

$$(a + b)^n = {_nC_0}\, a^n + {_nC_1}a^{n-1}b + {_nC_2}\, a^{n-2}b^2 + {_nC_3}^{n-3}b^3 + \ldots$$
$$\ldots + {_nC_{n-1}}a^1b^{n-1} + {_nC_n}b^n$$

where the numerical coefficients of the $n + 1$ terms are represented in combinatorial notation.

In the expansion formula for $(a + b)^n$:

- since ${_nC_0} = {_nC_n} = 1$, the first term is a^n and the last term is b^n;
- the coefficient of the kth term is ${_nC_{k-1}}$;
- the sum of the exponents in each term is n;
- in successive terms after the first, the power of a diminishes by 1, while the power of b increases by 1;

Example: To expand $(3x - y)^4$ evaluate the formula given by the binomial theorem for $n = 4$, $a = 3x$, and $b = -y$. Thus,

$$
\begin{aligned}
(a + b)^4 &= {_4C_0}a^4 + {_4C_1}a^3b \quad + {_4C_2}a^2b^2 \quad + {_4C_3}ab^3 \quad + {_4C_4}b^4 \\
&= 1a^4 \quad + 4a^3b \quad + 6a^2b^2 \quad + 4ab^3 \quad + 1b^4 \\
&= (3x)^4 + 4(3x)^3(-y) + 6(3x)^2(-y)^2 + 4(3x)(-y)^3 + (-y)^4 \\
&= 81x^4 - 108x^3y \quad + 54x^2y^2 \quad - 12xy^3 \quad + y^4
\end{aligned}
$$

Writing the kth term: Sometimes the only thing we need to know about the power of a binomial is what a particular term looks like. The kth term of the expansion of $(a + b)^n$ is given by the formula

$$k\text{th term of } (a + b)^n = {_nC_{k-1}}\, a^{n-(k-1)}\, b^{k-1}$$

where $k \leq n + 1$. For example, the expansion of $(x - 2y^3)^4$ contains $4 + 1 = 5$ terms. To find out what the middle or third term looks like, let $a = x$, $b = -2y$, $k = 3$, and $n = 4$. Thus,

Third term of $(x - 2y^3)^4 = {}_4C_{3-1}\,(x)^{4-(3-1)}\,(-2y^3)^{3-1}$

$$= {}_4C_2\,x^2\,(-2y^3)^2$$

$$= (6)\,(x^2)\,(4y^6)$$

$$= 24x^2y^6$$

KEY EXAMPLE

Find the numerical coefficient of the term that contains y^8x^5 in the expansion of $(y - 2x)^{13}$.

Solution: The first term contains y^{13}, the second term contains $y^{12}x$, the third term contains $y^{11}x^2$, and so forth. The sixth term contains y^8x^5.

$$\text{Sixth Term of } (y - 2x)^{13} = {}_{13}C_{6-1}\,y^{13-(6-1)}\,(-2x)^{6-1}$$

$$= {}_{13}C_5\,y^8\,(-2x)^5$$

$$= (1287)\,y^8\,(-32)x^5$$

$$= -41184\,y^8\,x^5$$

Hence, the numerical coefficient of the sixth term is -41184.

Theme 11 SEQUENCES, SERIES, AND INDUCTIONS

*L*ists of numbers sometimes follow a recognizable pattern. Some number patterns that involve only positive integers can be generalized by statements that look like formulas. These formulas are proved *indirectly* by using a special method of proof called **mathematical induction**.

Key 61 Arithmetic sequences and series

OVERVIEW *A sequence like 2, 5, 8, 11, 14, . . . is called an **arithmetic** sequence since each term after the first is obtained by adding a constant to the term that comes before it.*

Sequence (progression): A sequence is a list of numbers, called **terms**, written in a specific order. A **finite sequence** has a definite number of terms. An **infinite sequence** is nonending. An infinite sequence uses three trailing periods to indicate the pattern never ends, as in 2, 5, 8, 11, 14, . . .

Common difference: In an arithmetic sequence, subtracting any term from the term that follows it always results in the same number. This constant difference is called the **common difference** and is denoted by d. Thus, 2, 5, 8, 11, 14, . . . is an arithmetic sequence in which $d = 3$ since $5 - 2 = 8 - 5 = 11 - 8 = \ldots = 3$.

Means: The terms of a sequence that are between any two nonconsecutive terms are the means. In the arithmetic sequence 2, 5, 8, 11, 14, . . . , the numbers 5, 8, and 11 are the arithmetic means between 2 and 14.

Series: The indicated sum of the terms of a sequence forms a series.

Sequence formulas: If a_1 denotes the first term of an arithmetic sequence with n terms, a_n denotes the last term, and S_n represents the sum of all the terms from a_1 to a_n, then

- $a_n = a_1 + (n - 1)d$ \qquad [Formula 1]

- $S_n = \dfrac{n}{2}(a_1 + a_n)$ \qquad [Formula 2]

- $S_n = \dfrac{n}{2}[2a_1 + (n - 1)d]$ \quad [Formula 3]

KEY EXAMPLE

Insert three arithmetic means between -2 and 12.

Solution: We need to determine the three missing terms of an arithmetic sequence whose first term is -2 and whose last term is 12.

Substituting $n = 5$, $a_1 = -2$, and $a_5 = 12$ in formula 1 gives

$$12 = -2 + (5 - 1)\, d.$$

Solving this equation for d yields $d = 3.5$. Hence, the missing terms are 1.5, 5, and 8.5.

KEY EXAMPLE

If the eighth term of an arithmetic sequence is -18 and the third term is 7, find the sum of the first 30 terms.

Solution: There are six terms in the arithmetic subsequence in which $a_3 = 7$ is the first term and $a_8 = -18$ is the last term. By formula 1

$$-18 = 7 + (6 - 1)\, d.$$

Solving for d yields $d = -5$. Since $d = -5$ and $a_3 = 7$, $a_2 = 12$ and $a_1 = 17$. Now use formula 3 to find the sum:

$$S_{30} = \frac{30}{2}[2(17) + 29(-5)] = 1665$$

KEY EXAMPLE

How many numbers between 15 and 633 are divisible by 3? What is the sum of these numbers?

Solution: The numbers between 15 and 633 that are divisible by 3 form the arithmetic sequence 15, 18, 21, . . . , 633. Thus, $a_1 = 15$, $a_n = 633$, and $d = 3$. By formula 1,

$$633 = 15 + (n - 1)\, 3.$$

Solving for n gives n = 207. Hence, 207 numbers between 15 and 633 are divisible by 3. To find their sum, use formula 2 letting $a_1 = 15$, $a_n = 633$, and $n = 207$. Thus,

$$S_{207} = \frac{207}{2}(15 + 633) = 67,068.$$

Key 62   Geometric sequences and series

OVERVIEW   *A sequence like 2, 8, 32, 128, . . . is called a **geometric** sequence since each term after the first is obtained by multiplying the term that comes before it by a nonzero constant different than 1.*

Common ratio: In a geometric sequence dividing any term after the first by the term that comes before it gives the same nonzero number. This constant quotient is called the **common ratio** and is denoted by r. Thus, 2, 8, 32, 128, . . . is a geometric sequence in which $r = 4$ since

$$\frac{8}{2} = \frac{32}{8} = \frac{128}{32} = \cdots = 4.$$

Formulas: If S_n represents the sum of all the terms of a geometric sequence between a_1 and a_n where $r \neq 1$, then

- $a_n = a_1 r^{n-1}$ [Formula 4]

- $S_n = \dfrac{a_1(1 - r^n)}{1 - r}$ [Formula 5]

- $S_n = \dfrac{a_1 - a_n r}{1 - r}$ [Formula 6]

If the common ratio of an infinite geometric sequence is between -1 and 1, then the sum of its terms converges to a real number that can be found using the formula

- $S = \dfrac{a_1}{1 - r}$, where $|r| < 1$.   [Formula 7]

KEY EXAMPLE

Find the three geometric means between 36 and $\dfrac{64}{9}$, assuming $0 < r < 1$. Then find the sum of these five terms.

Solution: Since $0 < r < 1$, the terms of the geometric sequence are decreasing in value. Find r using formula 4. Substituting $n = 5$, $a_1 = 36$, and $a_5 = \dfrac{64}{9}$ into formula 4 yields

$$\frac{64}{9} = 36r^{5-1}.$$

Thus,

$$r^4 = \frac{16}{81} \text{ so } r = \frac{2}{3}.$$

Hence, the three geometric means are $a_2 = \frac{2}{3}(36) = 24$, $a_3 = \frac{2}{3}(24) = 16$,

and $a_4 = \frac{2}{3}(16) = \frac{32}{3}$. To find the sum of these five terms, use

formula 6 letting $a_1 = 36$, $a_5 = \frac{64}{9}$, $r = \frac{2}{3}$, and $n = 5$. Thus,

$$S_5 = \frac{36 - \left(\frac{64}{9}\right)\left(\frac{2}{3}\right)}{1 - \frac{2}{3}} = \frac{36 - \frac{128}{27}}{\frac{1}{3}}$$

$$= \frac{27\left(36 - \frac{128}{27}\right)}{27\left(\frac{1}{3}\right)} = \frac{972 - 128}{9} = \frac{844}{9}$$

KEY EXAMPLE

Express the repeating decimal 0.131313. . . as the ratio of two integers.

Solution: Since 0.131313 . . . = 0.13 + 0.13(0.01) + 0.13(0.01)^2 + . . . ,
the repeating decimal 0.131313 . . . can be written as the sum of an
infinite geometric sequence. Use formula 7 letting $a_1 = 0.13$ and
$r = .01$. Thus,

$$0.131313 \cdots = \frac{0.13}{1 - 0.01} = \frac{0.13}{0.99} = \frac{13}{99}$$

Key 63 Generalized sequences

OVERVIEW *Rather than listing the first few terms of a sequence and hoping that the pattern is apparent, a sequence can be described using a formula-type of expression.*

Sequence (function): A sequence is a function whose *domain* is a set of consecutive whole numbers that represent the position numbers of the terms in the sequence and whose *range* values are the actual **terms** of the sequence.

Notation: Instead of using the standard function notation $a(1)$, $a(2)$, $a(3)$, . . . , and so forth, the terms or function values of a sequence may be represented using the subscripted variables $a_1, a_2, a_3, \ldots, a_n, \ldots$ where the position of a term is given by its subscript. The notation $\{a_n\}$ refers to the entire sequence in which a_n is the nth term.

Example: In the sequence 10, 15, 20, and 25, the domain is the set of four position numbers, $\{1, 2, 3, 4\}$, and the range is $\{10, 15, 20, 25\}$. If a represents this sequence function, then $a_1 = 10$, $a_2 = 15$, $a_3 = 20$, and $a_4 = 25$.

n*th* Partial sum: The nth partial sum is the sum of the first n terms of a sequence, denoted by S_n. Thus,

$$S_n = a_1 + a_2 + a_3 + \ldots + a_n$$

Explicit formulas: A sequence may be defined by an explicit formula that relates the general term a_n to its subscript n, as in $a_n = 2n + 1$. Thus, $a_1 = 2(1) + 1 = 3$, $a_2 = 2(2) + 1 = 5$, $a_3 = 2(3) + 1 = 7$, and so forth.

Recursion formula: A recursion formula is one that relates a general term of a sequence to one or more terms that come before it. In the sequence 2, 4, 10, 28, . . . each term after the first is obtained by multiplying the previous term by 3 and then subtracting 2 from the product. The *recursion* formula that expresses this number pattern is $a_n = 3 (a_{n-1}) - 2$. For an arithmetic sequence whose common difference is d, the recursion formula is

$$a_n = a_{n-1} + d.$$

The recursion formula for a geometric sequence whose common ratio is r is

$$a_n = r\, a_{n-1}.$$

Sequences defined recursively: The recursion formula $a_n = 3 \, (a_{n-1}) - 2$ does not by itself provide enough information to obtain each term of the sequence it defines. If we also know that $a_1 = 2$, we can obtain any term by direct substitution:

$$a_2 = 3 \, (a_1) - 2 = 3(2) - 2 = 4$$

$$a_3 = 3 \, (a_2) - 2 = 3(4) - 2 = 10$$

$$a_4 = 3 \, (a_3) - 2 = 3(10) - 2 = 28$$

$$\vdots$$

A sequence is *defined recursively* when all of its terms can be computed from a recursion formula accompanied by the values of one or more given terms.

Fibonacci sequence: A Fibonacci sequence is the sequence 1, 1, 2, 3, 5, 8, 13, . . . in which each term after the second is formed by adding the two previous terms. The sequence can be defined recursively by $a_n = a_{n-1} + a_{n-2}$ with $a_1 = a_2 = 1$.

Disadvantage of recursive formulas: Finding terms whose subscripts are large may require a great deal of tedious computation since each of the terms that come before it also need to be calculated.

Key 64 Summation notation

OVERVIEW *The sum of any number of consecutive terms of a sequence can be indicated using the symbol Σ which is the capital Greek letter sigma.*

Sigma notation: The notation $\displaystyle\sum_{i=1}^{n} a_i$ represents the sum of the numbers a_i as i takes on consecutive integer values from i equals 1 (the value at the bottom of sigma) to i equals n (the value at the top of sigma). Thus,

$$\sum_{i=1}^{n} a_i = a_1 + a_2 + a_3 + \cdots + a_n$$

The sum $a_1 + a_2 + a_3 + \ldots + a_n$ is called the **expanded form** of $\displaystyle\sum_{i=1}^{n} a_i$ and variable i is called the **index variable** of this sum.

Example 1: The arithmetic series $3 + 6 + 9 + 12 + 15 + 18$ can be represented as $\displaystyle\sum_{k=1}^{6} 3k$.

Example 2: The geometric series $2 + 4 + 8 + 16 + 32$ can be represented as $\displaystyle\sum_{k=1}^{5} 2^k$.

Facts about summation notation:

- Although the index variable is always a nonnegative integer, it need not begin at 1. Thus,

$$\sum_{i=3}^{5} a_i = a_3 + a_4 + a_5.$$

- If a constant factor appears inside a summation sign, then it may be "passed through it." Thus,

$$\sum_{i=3}^{5} 2a_i = 2\sum_{i=3}^{5} a_i = 2(a_3 + a_4 + a_5).$$

- The terms that are being summed may be expressed in terms of the index variable. Thus,

$$\sum_{k=1}^{5} k^2 = 1^2 + 2^2 + 3^2 + 4^2 + 5^2 = 55.$$

- $\sum_{i=1}^{n} c = nc$. For example, $\sum_{i=1}^{4} 3 = 4 \cdot 3 = 12$.

- $\sum_{i=1}^{n} (a_i \pm b_i) = \sum_{i=1}^{n} a_i \pm \sum_{i=1}^{n} b_i$.

KEY EXAMPLE

Represent each sum using sigma notation:

(a) $2^3 + 3^3 + 4^3 + \ldots + 10^3$

(b) $(-1) + (4) + (-9) + (16) + (-25) + \ldots + n^2$

Solution: (a) $\sum_{k=2}^{10} k^3 = 2^3 + 3^3 + 4^3 + \cdots + 10^3$

(b) $\sum_{k=1}^{n} (-1)^k k^2 = -1 + 4 - 9 + 16 - 25 + \cdots + n^2$

KEY EXAMPLE

Evaluate $\sum_{k=1}^{3} (2k - 1)^2$.

Solution: Add terms having the form $(2k - 1)^2$ for $k = 1, 2,$ and 3.

$$\sum_{k=1}^{3} (2k - 1)^2 = [(2 \cdot 1 - 1)^2 + (2 \cdot 2 - 1)^2 + (2 \cdot 3 - 1)^2]$$

$$= 1^2 + 3^2 + 5^2 = 35$$

KEY EXAMPLE

Evaluate $\displaystyle\sum_{k=1}^{\infty}\left(\frac{4}{7}\right)^{k}$.

Solution: The given sum represents an infinite geometric series whose first term is $\frac{4}{7}$ and whose common ratio is $\frac{4}{7}$. Since $\frac{4}{7} < 1$, formula 7 (page 172) can be used. Thus,

$$\sum_{k=1}^{\infty}\left(\frac{4}{7}\right)^{k} = \frac{\dfrac{4}{7}}{1-\dfrac{4}{7}} = \frac{4}{3}.$$

KEY EXAMPLE

Evaluate: (a) $\displaystyle\sum_{k=2}^{6}\left(\frac{3}{2}\right)^{k}$ (b) $\displaystyle\sum_{k=2}^{\infty}\left(\frac{3}{2}\right)^{k}$

Solution: (a) The given sum represents a geometric series with $n = 5$, $a_1 = \left(\frac{3}{2}\right)^2 = \frac{9}{4}$ and $r = \frac{3}{2}$. Applying formula 5 (page 172) gives

$$\sum_{k=2}^{6}\left(\frac{3}{2}\right)^{k} = \frac{\left(\dfrac{9}{4}\right)\left(1-\left(\dfrac{3}{2}\right)^{5}\right)}{1-\dfrac{3}{2}} = \frac{\left(\dfrac{9}{4}\right)\left(1-\dfrac{243}{32}\right)}{-\dfrac{1}{2}} = \frac{\left(\dfrac{-1899}{128}\right)}{-\dfrac{1}{2}} = \frac{1899}{64}.$$

(b) Since the common ratio is greater than 1, the sum gets infinitely large and cannot be computed.

Key 65   Mathematical induction (part I)

OVERVIEW   *An indirect method of proof called **mathematical induction** is often used to prove formula type of statements which depend on a positive integer* n.

The need for induction: Let P_n represent a statement expressed in terms of a positive integer n like

$$1 + 3 + 5 + 7 + \ldots + (2n - 1) = n^2.$$

Statement P_n is either true or false. The truth values of statements P_1, $P_2, P_3,$ and P_4 can be determined by substituting 1, 2, 3, and 4 for n in the given statement which gives

$P_1:$	$1 = 1^2$	$P_3:$	$1 + 3 + 5 = 3^2$
$P_2:$	$1 + 3 = 2^2$	$P_4:$	$1 + 3 + 5 + 7 = 4^2$

Although statements $P_1, P_2, P_3,$ and P_4 are true, we cannot assume that this means that P_n will continue to hold true for each successive integer replacement for n. Since it is not practical or possible to test P_n for every positive integer value of n, we prove the statement is always true using an indirect method of proof called **mathematical induction**.

The "domino" principle of induction: Consider an endless row of standing dominoes in which P_1 is written on the first domino, P_2 is written on the second domino, and so forth. All of the dominoes falling in succession by one pushing over the domino that comes after it means that statements P$_1$, $P_2, \ldots, P_n$ are all true. For this to happen, two conditions must be met:

1. The first domino falls;
2. If the kth domino falls, it knocks down the $k + 1$st domino.

Mathematical principle of induction: To prove P_n is true for all positive integers n, complete these two steps:

Step 1. Show P_1 is true;

Step 2. Assume P_k is true, and show P_{k+1} is also true.

Step 1 is called the **verification step** and corresponds to showing that the first domino falls. Step 2 is called the **induction step** and P_k the **induction hypothesis**. The procedure in this step is analogous to showing that if the kth domino falls, it pushes down the $k + 1$st domino. A general strategy used in completing this step is to manipulate P_k algebraically in order to obtain P_{k+1}. This proves P_k *implies* P_{k+1}.

KEY EXAMPLE

Prove : $1 + 3 + 5 + 7 + . . . + (2n - 1) = n^2.$

Solution: Let P_n represent the given statement.

Step 1. Show P_1 is true. When $n = 1$, the left side of the statement is the first term, 1. The right side is $1^2 = 1$. Since $1 = 1$, P_1 is true.

Step 2. In the given statement, replace n by k to form P_k, and replace n by $k + 1$ to form P_{k+1}. Thus,

$$P_{k:}\quad 1 + 3 + 5 + 7 + . . . + (2k - 1) \qquad = k^2$$

$$P_{k+1}:\ 1 + 3 + 5 + 7 + . . . + (2k - 1) + (2k + 1) = (k + 1)^2$$

Assume P_k is true. Try to transform the equation represented by P_k into an equivalent equation that looks exactly like the equation represented by P_{k+1}.

Comparing P_k to P_{k+1}, we see that the left side of the equation represented by P_{k+1} contains an additional term, $(2k + 1)$. This suggests that we add $(2k + 1)$ to both sides of the equation represented by P_k :

$$1 + 3 + 5 + 7 + . . . + (2k - 1) + (2k + 1) = k^2 + (2k + 1)$$

Simplify the right side of the equation by factoring $k^2 + 2k + 1$ as $(k + 1)^2$. Thus,

$$1 + 3 + 5 + 7 + . . . + (2k - 1) + (2k + 1) = (k + 1)^2.$$

Since the previous statement is P_{k+1}, and was obtained directly from P_k, the truth of P_k implies the truth of P_{k+1}. Hence, P_n is true for every positive integer n.

KEY EXAMPLE

Prove: $1 + 2 + 4 + 8 + . . . + 2^{n-1} = 2^n - 1$

Solution: Let P_n represent the given statement.

Step 1. Statement P_1 is $2^{1-1} = 2^1 - 1$. Since each side of the equation is equal to 1, P_1 is true.

Step 2. Form P_k and P_{k+1} from P_n by replacing n with k and then with $k + 1$:

$$P_k:\ 1 + 2 + 4 + 8 + . . . + 2^{k-1} \qquad = 2^k - 1$$

$$P_{k+1}:\ 1 + 2 + 4 + 8 + . . . + 2^{k-1} + 2^k = 2^{k+1} - 1$$

Obtain P_{k+1} from P_k by adding 2^k to both sides of the equation represented by statement P_k:

$$1 + 2 + 4 + 8 + \ldots + 2^{k-1} + 2^k$$

$$= (2^k - 1) + 2^k$$

$$= 2 \cdot 2^k - 1$$

$$= 2^{k+1} - 1$$

The last formula statement corresponds to P_{k+1} which was obtained directly from P_k. Thus, the truth of P_k implies the truth of P_{k+1}, and the proof by mathematical induction is complete.

Key 66 Mathematical induction (part II)

OVERVIEW *Mathematical induction is used in proofs other than those that establish general sum formulas.*

Proving divisibility: Some algebraic expressions that depend on n are always divisible by the same constant.

Example: The expression $4^n - 1$ is divisible by 3 for *all* positive integer value replacements of n. We can prove this statement using mathematical induction.

Step 1. For $n = 1$, $4^1 - 1 = 3$ which is divisible by 3.

Step 2. Assume P_k: "$4^k - 1$ is divisible by 3" is true and show that P_{k+1}: "$4^{k+1} - 1$ is divisible by 3" is also true.

If $4^k - 1$ is divisible by 3, $4^k - 1$ must be some integer multiple of 3. Hence,

$$4^k - 1 = 3p,$$

where p is some positive integer. Thus,

$$4^k = 3p + 1.$$

Multiplying each side of this equation by 4 and then subtracting 1 from each side gives

$$4^{k+1} - 1 = 3(4p + 1).$$

Since $4p + 1$ is an integer, $4^{k+1} - 1$ equals an integer multiple of 3 and is, therefore, divisible by 3. This establishes P_{k+1} so the proof is complete.

Proving a general inequality statement: Some pairs of algebraic expressions that depend on n always have the same size order.

Example: The statement P_n: $n! > 2^n$ is true for all integer replacements for n provided $n \geq 4$. We can prove this inequality statement using mathematical induction.

Step 1. The initial value of n for which the formula is defined, called the **anchor value**, is 4. The verification step must show that P_4 is true where P_4 represents $4! > 2^4$. Since $4! = 24$ and $2^4 = 16$, P_4 is true.

Step 2. Assume P_k: $k! > 2^k$ is true, and show that P_{k+1}: $(k + 1)! > 2^{k+1}$ is also true. Multiplying both sides of the inequality represented by P_k by $(k+1)$ gives

$$(k+1) \, ! \; > \; (k + 1) \, 2^k.$$

Since $k \geq 4$, $k + 1 > 2$ so

$$(k+1)! \; > \; (k + 1) \, 2^k \; > \; (2) \, 2^k.$$

If $A > B > C$, then $A > C$. From this property it follows that

$$(k + 1) \, ! \; > \; 2^{k+1}$$

which shows P_k implies the truth of P_{k+1} and, as a result, completes the proof.

Theme 12 LINEAR AND
NONLINEAR SYSTEMS

*T*ranslating the conditions of a problem situation into mathematical language may lead to a **system** of two or more equations in two or more different variables. Solving a system of equations means finding the values of the variables that make all of the equations of the system true at the same time. This Key includes some methods for representing and solving systems of equations that involve two or more variables.

Key 67   Solving systems of two linear

equations

OVERVIEW   *Any common solution to a pair of linear equations in two variables can be obtained algebraically or graphically.*

System of linear equations in two variables: A collection of equations in which each equation can be put into the **standard form** $ax + by = c$, where a, b, and c are constants is called a system of linear equations. The solution set of a linear system in two variables is the set of all ordered pairs, if any, that simultaneously makes all of the equations in the system true. The solution set of the system shown at the right is the ordered pair $(6, -1)$ since $x = 6$ and $y = -1$ satisfies both equations.

$$x + y = 5$$
$$x - y = 7$$

To solve a two-equation linear system by substitution: Solve one of the equations for one of the variables in terms of the other. Then substitute the expression obtained for that variable in the other equation.

Example: In the system shown at the right, solving for y in equation (ii) gives $y = 2x - 26$. Replacing y with $2x - 26$ in equation (i) produces an equation in one variable,

$$3x - 2y = 42 \quad \text{(i)}$$
$$2x - y = 26 \quad \text{(ii)}$$

$$3x - 2(2x - 26) = 42,$$

whose solution is $x = 10$. The corresponding value of y can be obtained by replacing x with 10 in either equation (i) or equation (ii). Using equation (ii), we obtain

$$2(10) - y = 26$$

so

$$y = -6.$$

The solution is the ordered pair $(10, -6)$. In order to check that this is the correct ordered pair, you must verify that $x = 10$ and $y = -6$ satisfies both equation (i) and equation (ii). The check is left for you.

To solve a two-equation linear system by elimination: Write each equation in standard form, and then compare the coefficients of like variables. If necessary, multiply one or both equations by an appropriate constant so that in the resulting pair of equations one of the variables has numerical coefficients that are additive inverses. Eliminate this variable by adding corresponding sides of the two equations.

Example: In the system shown at the right, variable x can be eliminated by multiplying equation (i) by 5, multiplying equation (ii) by -3, and then adding corresponding sides of the resulting equations. Thus,

$$3x + 4y = 9 \qquad \text{(i)}$$

$$5x + 6y = 13 \qquad \text{(ii)}$$

$3x + 4y = 9$	*multiply by 5*	$15x + 20y = 45$	(I)
$5x + 6y = 13$	*multiply by -3*	$-15x - 18y = -39$	(II)
		$2y = 6$	(I) + (II)

Since $y = 3$, replacing y with this value in either equation gives $x = -1$. Alternatively, variable y could have been eliminated by multiplying equation (i) by 3, multiplying equation (ii) by -2, and then adding the corresponding sides of the resulting pair of equations.

To solve a two-equation linear system by graphing: Graph each line on the same set of coordinate axes. Any point of intersection of these lines is a solution of the system. There are three possibilities.

1. The lines intersect at exactly one point so the solution is unique. In this case the system is **consistent**.
2. The lines are parallel so there is no solution. In this case the system is **inconsistent.**
3. The lines coincide so there are an infinite number of solution points. In this case one equation is a multiple of the other equation and the system is **dependent**.

Key 68 Matrices

OVERVIEW *A rectangular table of numbers that are arranged in* m *horizontal rows and* n *vertical columns forms an* m × n *(read as "*m *by* n*")* **matrix**. *The plural of* **matrix** *is "matrices."*

Matrix notation: A matrix is usually named by a boldfaced capital letter and its terms, called **elements**, are enclosed by brackets.

$$
\begin{array}{c}
\textit{Columns} \\
\begin{array}{cccc}
1 & 2 & 3 & 4 \\
\downarrow & \downarrow & \downarrow & \downarrow
\end{array} \\
\mathbf{A} = \begin{bmatrix} 2 & -5 & 6 & 9 \\ 8 & 1 & 7 & 3 \\ -3 & -4 & 0 & \sqrt{5} \end{bmatrix}
\begin{array}{l}
\rightarrow \; \textit{Row 1} \\
\rightarrow \; \textit{Row 2} \\
\rightarrow \; \textit{Row 3}
\end{array}
\end{array}
$$

Matrix **A** is a 3 × 4 matrix since it has three rows and four columns. Thus, the **dimension** or size of this matrix is 3 × 4.

Locating an element of a matrix: To refer to an element of a matrix, give its row number followed by its column number. The element that occupies the ith row and jth column in matrix **A** is denoted by the double-subscripted variable, a_{ij}. For instance, $a_{23} = 7$ and $a_{32} = -4$. Similarly, the element in row i and column j of **B** is denoted by b_{ij}. Sometimes the bracketed notation $[a_{ij}]$ is used to represent the entire array of elements of matrix **A**.

Square matrix: A matrix that has the same number of columns as rows. The **main diagonal** of a square matrix is the line that contains all elements whose row and column numbers are the same ($i = j$), as shown for the 3 × 3 matrix.

$$
\begin{bmatrix} a_{11} & a_{12} & a_{13} \\ a_{21} & a_{22} & a_{23} \\ a_{31} & a_{32} & a_{33} \end{bmatrix}
$$

Matrix addition and subtraction: Matrices that have the same dimension are added or subtracted by combining corresponding pairs of elements. Thus,

$$\begin{bmatrix} 5 & 7 \\ -2 & 4 \end{bmatrix} + \begin{bmatrix} 1 & -4 \\ 2 & 3 \end{bmatrix} = \begin{bmatrix} 5+1 & 7+(-4) \\ -2+2 & 4+3 \end{bmatrix} = \begin{bmatrix} 6 & 3 \\ 0 & 7 \end{bmatrix}$$

The commutative and associative properties hold for matrix addition.

Scalar multiplication: To multiply a matrix by a number, multiply each element of the matrix by that number. Thus, $k A = k [a_{ij}] = [k a_{ij}]$ where k is some real number called the **scalar**. For instance,

$$2\begin{bmatrix} -1 & 3 \\ 4 & 0 \end{bmatrix} = \begin{bmatrix} 2(-1) & 2(3) \\ 2(4) & 2(0) \end{bmatrix} = \begin{bmatrix} -2 & 6 \\ 8 & 0 \end{bmatrix}$$

Row by column multiplication: Multiplying a row of a matrix by a column of another matrix is accomplished by multiplying the first element in the row by the first element in the column, the second element in the row by the second element in the column, and so forth, and then adding these products. For instance,

$$\begin{bmatrix} 2 & 4 & -1 \end{bmatrix} \begin{bmatrix} 5 \\ 3 \\ 8 \end{bmatrix} = [2 \cdot 5 + 4 \cdot 3 + (-1)8] = [14]$$

Matrix multiplication: Multiplying the ith row of matrix **A** by the jth column of matrix **B** gives the element in the ith row and jth column of their product matrix. For example,

$$\mathbf{AB} = \begin{bmatrix} 5 & -3 & 1 \\ -1 & 0 & 4 \end{bmatrix} \begin{bmatrix} 6 & 1 \\ -2 & 3 \\ 7 & -5 \end{bmatrix} = \begin{bmatrix} \text{row 1 of } \mathbf{A} \text{ times} & \text{row 1 of } \mathbf{A} \text{ times} \\ \text{column 1 of } \mathbf{B} & \text{column 2 of } \mathbf{B} \\ \text{row 2 of } \mathbf{A} \text{ times} & \text{row 2 of } \mathbf{A} \text{ times} \\ \text{column 1 of } \mathbf{B} & \text{column 2 of } \mathbf{B} \end{bmatrix}$$

$$= \begin{bmatrix} 5(6) + (-3)(-2) + 1(7) & 5(1) + (-3)(3) + 1(-5) \\ (-1)6 + 0(-2) + 4(7) & (-1)(1) + 0(3) + 4(-5) \end{bmatrix}$$

$$= \begin{bmatrix} 43 & -9 \\ 22 & -21 \end{bmatrix}$$

Compatibility of matrices: In order to be able to perform row by column matrix multiplication, the number of elements in each row of the first matrix must equal the number of elements in each column of second matrix. In general, matrices **A** and **B** are **compatible** for multiplication only if the number of columns of matrix **A** equals the number of rows of matrix **B**. If two matrices are not compatible, then they cannot be multiplied.

Dimension of a product matrix: If **A** is an $m \times n$ matrix and **B** is an $n \times p$ matrix, then **A** and **B** are compatible matrices and the product matrix **AB** is an $m \times p$ matrix. Thus, **AB** has the same number of rows as **A** and the same number of columns as **B**.

AB ≠ BA: The order in which two matrices are multiplied matters. The commutative property does *not* hold for matrix multiplication, although the associative and distributive properties do.

Identity matrix for multiplication: A square $n \times n$ matrix, denoted by $\mathbf{I}_n$, that has a 1 in each position along its main diagonal and a 0 in every other position is an identity matrix. Thus,

$$\mathbf{I}_2 = \begin{bmatrix} 1 & 0 \\ 0 & 1 \end{bmatrix} \quad \text{and} \quad \mathbf{I}_3 = \begin{bmatrix} 1 & 0 & 0 \\ 0 & 1 & 0 \\ 0 & 0 & 1 \end{bmatrix}$$

The identity matrix plays the same role in matrix multiplication as 1 does in arithmetic multiplication. Thus, if **A** is a square matrix, then

$$\mathbf{A} \ \mathbf{I} = \mathbf{I} \ \mathbf{A} = \mathbf{A}.$$

The subscript n in $\mathbf{I}_n$ is omitted since we always assume it has the same size as the matrix it multiplies.

Key 69 Determinants

OVERVIEW *Every* n × n *square matrix* **A** *has an nth order* **determinant**, *denoted by* |**A**|, *that associates a numerical value with the matrix.*

Second-order determinant: If **A** is the 2 × 2 square matrix $\begin{bmatrix} a_{11} & a_{12} \\ a_{21} & a_{22} \end{bmatrix}$,

then |**A**| denotes a second-order determinant that is defined as the product of the elements along the main diagonal minus the product of the elements along the other diagonal. That is,

$$|\mathbf{A}| = \begin{vmatrix} a_{11} & a_{12} \\ a_{21} & a_{22} \end{vmatrix} = (a_{11} \cdot a_{22}) - (a_{21} \cdot a_{12})$$

For example,

$$\begin{vmatrix} 8 & 2 \\ 7 & 3 \end{vmatrix} = (8 \cdot 3) - (7 \cdot 2) = 24 - 14 = 10$$

Minor of element a_{ij} of an nth order determinant: The determinant of order $n - 1$, denoted by M_{ij}, formed by deleting the row and column that contains a_{ij}, is the minor of element a_{ij}. For the third order determinant given at the right, the minor of 4 $(= a_{12})$ is the second order determinant, denoted by M_{12}, obtained by crossing out the row and column that contains 4 in the original determinant. Thus,

$$\begin{vmatrix} 3 & 4 & -2 \\ 2 & -1 & -3 \\ 5 & 1 & 0 \end{vmatrix}$$

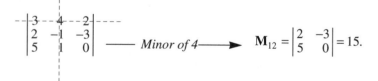

Cofactor of element a_{ij} of an nth order determinant: This value is denoted by $\mathbf{C}_{ij}$ where $\mathbf{C}_{ij} = (-1)^{i+j} \mathbf{M}_{ij}$. If

$$\mathbf{M}_{12} = 15,$$

then

$$\mathbf{C}_{ij} = (-1)^{1+2} (15) = -15.$$

To evaluate a third-order determinant: Write the determinant in terms of second-order determinants as follows:

$$\begin{vmatrix} a_{11} & a_{12} & a_{13} \\ a_{21} & a_{22} & a_{23} \\ a_{31} & a_{32} & a_{33} \end{vmatrix} = a_{11}\begin{vmatrix} a_{22} & a_{23} \\ a_{32} & a_{33} \end{vmatrix} - a_{21}\begin{vmatrix} a_{12} & a_{13} \\ a_{32} & a_{33} \end{vmatrix} + a_{31}\begin{vmatrix} a_{12} & a_{13} \\ a_{22} & a_{23} \end{vmatrix}$$

Notice on the right side of the equation, the expression that multiplies a_{11} is the cofactor of a_{11}, the expression that multiplies a_{21} (including the negative sign) is the cofactor of a_{21}, and the expression that multiplies a_{31} is the cofactor of a_{31}. Thus,

$$\begin{vmatrix} a_{11} & a_{12} & a_{13} \\ a_{21} & a_{22} & a_{23} \\ a_{31} & a_{32} & a_{33} \end{vmatrix} = a_{11}C_{11} + a_{21}C_{21} + a_{31}C_{33}.$$

The previous equation gives the expansion of a third-order determinant along its *first* row since the right side of the equation represents the sum of the products of each element in the first row and its cofactor. A determinant of any order may be expanded by cofactors along *any* row or column. If we choose to expand a third-order determinant A along its second column, then,

$$|\mathbf{A}| = a_{12}C_{12} + a_{22}C_{22} + a_{32}C_{32}$$

KEY EXAMPLE

Evaluate $\quad |\mathbf{A}| = \begin{vmatrix} 3 & 4 & -2 \\ 2 & -1 & -3 \\ 5 & 1 & 0 \end{vmatrix}$

Solution: Evaluate the determinant by choosing any convenient row or column of the determinant and expanding by cofactors along this row or column. Expand along the third column since it contains a 0. Thus,

$$\begin{aligned}
|\mathbf{A}| &= \quad (-2)C_{13} \quad + \quad (-3)C_{23} \quad + \quad 0C_{33} \\
&= (-2)(-1)^4\begin{vmatrix} 2 & -1 \\ 5 & 1 \end{vmatrix} + (-3)(-1)^5\begin{vmatrix} 3 & 4 \\ 5 & 1 \end{vmatrix} + \quad 0 \\
&= \quad (-2)(7) \quad + \quad (3)(-17) \quad + \quad 0 \\
&= \quad -14 + (-51) \\
&= \quad -65
\end{aligned}$$

Determinants that evaluate to zero: A determinant will evaluate to 0 if it has any of the following properties:

- All the elements of any row or column are 0;
- Two rows or two columns are identical;
- A row (or column) is a multiple of another row (or column).

Operations that do not affect a determinant:
- Interchanging all the rows and columns;
- Replacing any row (or column) with the sum of that row (or column) and a fixed multiple of another row (or column).

Operations that affect a determinant:
- Interchanging any two rows or any two columns of a determinant changes the sign of the determinant;
- Multiplying each element in a row (or column) by some constant multiplies the value of the determinant by the same constant.

Key 70 Cramer's rule

OVERVIEW *Systems of linear equations can be solved using determinants.*

Cramer's rule for systems of two linear equations: If the system at the right has exactly one solution, then the solution is given by

$$a_1x + b_1y = c_1$$
$$a_2x + b_2y = c_2$$

$$x = \frac{D_x}{D} \quad \text{and} \quad y = \frac{D_y}{D}$$

where

$$D = \begin{vmatrix} a_1 & b_1 \\ a_2 & b_2 \end{vmatrix} \neq 0, \quad D_x = \begin{vmatrix} c_1 & b_1 \\ c_2 & b_2 \end{vmatrix}, \quad \text{and} \quad D_y = \begin{vmatrix} a_1 & c_1 \\ a_2 & c_2 \end{vmatrix}$$

Notice that D is the determinant whose members are the coefficients of the variables in the given system of equations, D_x is the determinant obtained from the determinant D by replacing the coefficients of x with c_1 and c_2, and D_y is the determinant obtained from the determinant D by replacing the coefficients of y with c_1 and c_2.

If $D = 0$, then Cramer's rule cannot be used. In this case the system is either dependent or inconsistent. If $D = D_x = D_y = 0$, the system is dependent and has an infinite number of solutions. If $D = 0$, but at least one of D_x or D_y is *not* 0, then the system is inconsistent and there are no solutions. The following Key Example illustrates how Cramer's rule can be extended to solve a system of three linear equations.

KEY EXAMPLE

Use Cramer's Rule to solve the system of equations shown at the right.

$$x - 2y + z = 3$$
$$2x - z = -7$$
$$y + 2z = 5$$

Solution: The determinant D of the coefficients is $D = \begin{vmatrix} 1 & -2 & 1 \\ 2 & 0 & -1 \\ 0 & 1 & 2 \end{vmatrix}$.

Replacing each column of this determinant with the constant terms gives

$$D_x = \begin{vmatrix} 3 & -2 & 1 \\ -7 & 0 & -1 \\ 5 & 1 & 2 \end{vmatrix}, \quad D_y = \begin{vmatrix} 1 & 3 & 1 \\ 2 & -7 & -1 \\ 0 & 5 & 2 \end{vmatrix}, \quad \text{and} \quad D_z = \begin{vmatrix} 1 & -2 & 3 \\ 2 & 0 & -7 \\ 0 & 1 & 5 \end{vmatrix}$$

You can verify that $D = 11$, $D_x = -22$, $D_y = -11$, and $D_z = 33$. Thus,

$$x = \frac{D_x}{D} = \frac{-22}{11} = -2; \quad y = \frac{D_y}{D} = \frac{-11}{11} = -1; \quad \text{and} \quad z = \frac{D_z}{D} = \frac{33}{11} = 3.$$

Key 71 Triangularizing linear systems

OVERVIEW *A system of linear equations can be put into matrix form and then solved by arranging the elements of the matrix in a special triangular form.*

Triangular form of a three-equation linear system: A system of three equations in three unknowns, say x, y, and z, is in triangular form when variable z is the only variable in the last equation and variables y and z are the only variables in the middle equation.

Example: the system at the right is in triangular form.

$$\begin{aligned} 4x + 3y - z &= 1 \\ 2y + 3z &= 4 \\ -7z &= 14 \end{aligned}$$

Its solution can be obtained by solving the last equation for z so $z = -2$. Substituting this value of z in the middle equation and solving for y gives $y = 5$. Lastly, substituting the values for y and z in the first equation and solving for x gives $x = -4$. This method of solution is called **"back-substitution."**

Triangular form of a matrix: In this matrix all the elements below the main diagonal are 0, as shown below:

$$\begin{bmatrix} a_1 & a_2 & a_3 & a_4 \\ 0 & b_1 & b_2 & b_3 \\ 0 & 0 & c_1 & c_2 \\ 0 & 0 & 0 & d_1 \end{bmatrix}$$

Matrix form of a system of linear equations: To put a system of three equations in three unknowns in matrix form, write the first equation of the system in the form $Ax + By + Cz = k$. Then write the coefficients A, B, C, and the constant k for this equation as the elements of the first row of a matrix. Do the same for each of the other equations of the system. For example,

$$\begin{aligned} x + 2y + z &= 2 \\ 4x - y - 3z &= -4 \\ 2x + y - z &= -2 \end{aligned}$$

corresponds to

$$\begin{bmatrix} 1 & 2 & 1 & \vdots & 2 \\ 4 & -1 & -3 & \vdots & -4 \\ 2 & +1 & -1 & \vdots & -2 \end{bmatrix}$$

The 3×3 matrix to the left of the vertical dashed line is the **coefficient matrix.** The 3×4 matrix is called the **augmented matrix** since it includes the constant terms as well as the numerical coefficients of the variables. Each row of the augmented matrix corresponds to an equation of the system. If an equation of the system does not contain a particular variable, then its coefficient is 0 which is used as the corresponding element in the augmented matrix.

Row equivalent matrices: Row equivalent matrices are obtained from a given matrix by using one or more of the following **elementary row operations:**

1. Interchanging two rows.
2. Multiplying or dividing a row by a nonzero constant.
3. Replacing a row with the sum of that row and a multiple of another row.

A row equivalent matrix corresponds to a new system of linear equations that is equivalent to the system of equations that the original matrix represents.

Matrix solution of a linear system of equations: Form the augmented matrix that represents the given system of linear equations. Convert this matrix into a row equivalent matrix in triangular form. Then translate each row of the triangular matrix into a linear equation and solve the resulting system of equations by back-substitution.

KEY EXAMPLE

Solve the following system of equations:

$$\begin{aligned} x + 2y + z &= 2 \\ 4x - y - 3z &= -4 \\ 2x + y - z &= -2 \end{aligned}$$

Solution: Form the augmented matrix and then convert it into a row-equivalent matrix in *triangular form*. Replace the second row (R2) with the sum of that row (R2) and the third row (R3) multiplied by –2. Thus,

$$\begin{bmatrix} 1 & 2 & 1 & 2 \\ 4 & -1 & -3 & -4 \\ 2 & +1 & -1 & -2 \end{bmatrix} \quad \text{R2} \rightarrow +\text{R2} + (-2)\text{R3} \quad \begin{bmatrix} 1 & 2 & 1 & 2 \\ 0 & -3 & -1 & 0 \\ 2 & +1 & -1 & -2 \end{bmatrix}$$

Replace the third row (R3) by the sum of that row (R3) and the first row (R1) multiplied by –2 .

$$\begin{bmatrix} 1 & 2 & 1 & \vdots & 2 \\ 0 & -3 & -1 & \vdots & 0 \\ 2 & +1 & -1 & \vdots & -2 \end{bmatrix} \quad R3 \to R3+(-2)R1 \quad \begin{bmatrix} 1 & 2 & 1 & \vdots & 2 \\ 0 & -3 & -1 & \vdots & 0 \\ 0 & -3 & -3 & \vdots & -6 \end{bmatrix}$$

Replace the third row by the sum of that row and the second row multiplied by –1. The matrix that results will be in triangular form

$$\begin{bmatrix} 1 & 2 & 1 & \vdots & 2 \\ 0 & -3 & -1 & \vdots & 0 \\ 0 & -3 & -3 & \vdots & -6 \end{bmatrix} \quad R3 \to R3+(-1)R2 \quad \begin{bmatrix} 1 & 2 & 1 & \vdots & 2 \\ 0 & -3 & -1 & \vdots & 0 \\ 0 & 0 & -2 & \vdots & -6 \end{bmatrix}$$

Once the matrix is in triangular form, it can be converted back into an equivalent system of equations which can be solved by back-substitution. Solving the last equation we obtain $z = 3$.

$$\begin{aligned} x + 2y + z &= 2 \\ -3y - z &= 0 \\ -2z &= -6 \end{aligned}$$

Substituting into the second equation and solving for y gives $y = -1$. Substituting the solutions for y and z in the first equation and solving for x gives $x = 1$.

Key 72 Recognizing linear systems that have many or no solutions

OVERVIEW   *In the process of triangularizing a linear system of equations we may encounter a system that does not have exactly one solution.*

A linear system with infinitely many solutions: After triangularizing a linear system, one of its equations may have the form of 0 times a variable equals 0. We can arbitrarily assign this variable any value and then calculate the corresponding values of other variables.

Example: In the triangularized system of equations shown at the right, the last equation tells us that z can have *any* value. Solving the second equation for y gives $y = 5 - z$.

$$x + y - 2z = 9$$
$$y + z = 5$$
$$0z = 0$$

Substituting this expression for y in the first equation and solving for x gives $3z + 4$. Hence, the infinite set of solutions for the original system of equations can be represented as

$$x = 3z + 4, \quad y = 5 - z, \quad \text{and} \quad z = \text{any real number.}$$

For example, if $z = 1$, then $x = 3(1) + 4 = 7$, and $y = 5 - 1 = 4$.

A linear system with no solutions: Consider the same system of three equations from the previous example, but with the equation $0z = 0$ replaced by $y + z = -3$. The last equation now contradicts the second equation so the system is inconsistent and, as a result, has no solutions.

Key 73 Finding the inverse of a matrix

OVERVIEW *Since* $5 \times \dfrac{1}{5} = 1$, $\dfrac{1}{5}$ *is called the multiplicative inverse of 5. Similarly, if* $AB = I$, *then* B *is called the inverse of matrix* A. *The inverse of matrix* A *is usually written as* A^{-1}. *A matrix may or may not have an inverse.*

Inverse of a square matrix A: The square matrix denoted by A^{-1}, provided it exists, that has the property

$$A A^{-1} = A^{-1}A = I.$$

Nonsingular matrix: A square matrix that has an inverse matrix is nonsingular.

Finding the inverse of a matrix: If A is an $n \times n$ nonsingular matrix, then its inverse is obtained by following these steps:

1. Append to matrix A the identity matrix I_n so that the resulting $n \times 2n$ matrix has the form $[A \mid I_n]$.
2. Use elementary row operations to transform this matrix into a row equivalent matrix with the form $[I_n \mid B]$.
3. Form matrix B which is the inverse of A. That is, $B = A^{-1}$.

KEY EXAMPLE

Find the inverse of matrix A where $A = \begin{bmatrix} 1 & 2 \\ 3 & -1 \end{bmatrix}$.

Solution: Append the 2×2 identity matrix to matrix A:

$$\begin{bmatrix} 1 & 2 & \vdots & 1 & 0 \\ 3 & -1 & \vdots & 0 & 1 \end{bmatrix}$$

Replace the second row (R2) by the sum of that row (R2) and the first row (R1) multiplied by –3. Then replace the second row (R2) by that row multiplied by $-\dfrac{1}{7}$. Thus,

$$\begin{bmatrix} 1 & 2 & \vdots & 1 & 0 \\ 3 & -1 & \vdots & 0 & 1 \end{bmatrix} \xrightarrow{\begin{array}{c} R2 \to R2 + (-3)R1 \\ \\ R2 \to \left(-\dfrac{1}{7}\right)R2 \end{array}} \begin{bmatrix} 1 & 2 & \vdots & 1 & 0 \\ 0 & 1 & \vdots & \dfrac{3}{7} & \dfrac{-1}{7} \end{bmatrix}$$

Replace the first row (R1) by the sum of that row (R1) and the second row (R2) multiplied by –2 . Thus,

$$\begin{bmatrix} 1 & 2 & \vdots & 1 & 0 \\ 0 & 1 & \vdots & \dfrac{3}{7} & \dfrac{-1}{7} \end{bmatrix} \xrightarrow{R1 \to R1 + (-2)R2} \begin{bmatrix} 1 & 0 & \vdots & \dfrac{1}{7} & \dfrac{2}{7} \\ 0 & 1 & \vdots & \dfrac{3}{7} & \dfrac{-1}{7} \end{bmatrix}$$

Thus, $\mathbf{A}^{-1} = \begin{bmatrix} \dfrac{1}{7} & \dfrac{2}{7} \\ \dfrac{3}{7} & \dfrac{-1}{7} \end{bmatrix}$

Key 74   Solving linear systems
by matrix inversion

OVERVIEW   *In algebra, if* ax = b *then multiplying each side of the equation by the multiplicative inverse (reciprocal) of* a *gives the solution* $x = \dfrac{b}{a} = a^{-1}b$. *Similarly, if* **AX = K**, *then* **X = A⁻¹K**.

Representing a linear system as a matrix product: A system of linear equations can be represented by a *single* matrix equation. A linear system of equations having the form at the right may be written as the matrix equation **AX = K** where

$$a_1 x + b_1 y + c_1 z = k_1$$
$$a_2 x + b_2 y + c_2 z = k_2$$
$$a_3 x + b_3 y + c_3 z = k_3$$

$$\mathbf{A} = \begin{bmatrix} a_1 & b_1 & c_1 \\ a_2 & b_2 & c_2 \\ a_3 & b_3 & c_3 \end{bmatrix}, \quad \mathbf{X} = \begin{bmatrix} x \\ y \\ z \end{bmatrix}, \quad \text{and} \quad \mathbf{K} = \begin{bmatrix} k_1 \\ k_2 \\ k_3 \end{bmatrix}$$

Thus, the matrix equation **X = A⁻¹K** represents the solution to the original system of linear equations. The solution for any system of *n* linear equations in *n* unknowns, provided it exists, can be represented in this way. The following Key Example illustrates how to obtain the solution to a two-equation linear system using matrix inversion.

KEY EXAMPLE

Solve the system

$$x + 2y = -3$$
$$3x - y = 5$$

Solution: The given system of equations can be represented by the matrix equation

$$\mathbf{AX = K}$$

where

$$\mathbf{A} = \begin{bmatrix} 1 & 2 \\ 3 & -1 \end{bmatrix}$$

$$\mathbf{X} = \begin{bmatrix} x \\ y \end{bmatrix}$$

and

$$\mathbf{K} = \begin{bmatrix} -3 \\ 5 \end{bmatrix}.$$

From the example in Key 73 (page 199),

$$\mathbf{A}^{-1} = \begin{bmatrix} \dfrac{1}{7} & \dfrac{2}{7} \\ \dfrac{3}{7} & \dfrac{-1}{7} \end{bmatrix}$$

Since $\mathbf{X} = \mathbf{A}^{-1}\mathbf{K}$,

$$\mathbf{X} = \begin{bmatrix} \dfrac{1}{7} & \dfrac{2}{7} \\ \dfrac{3}{7} & \dfrac{-1}{7} \end{bmatrix} \begin{bmatrix} -3 \\ 5 \end{bmatrix} = \begin{bmatrix} \dfrac{-3}{7} + \dfrac{10}{7} \\ \dfrac{-9}{7} + \dfrac{-5}{7} \end{bmatrix} = \begin{bmatrix} 1 \\ -2 \end{bmatrix} = \begin{bmatrix} x \\ y \end{bmatrix}.$$

Hence, $x = 1$ and $y = -2$.

Key 75   Partial fractions

OVERVIEW   *Sometimes, as in the study of calculus, it is necessary to start with a fraction like the one on the right side of the equation* $\dfrac{2}{x} + \dfrac{5}{x-3} = \dfrac{7x-6}{x^2-3x}$ *and to **decompose** it into simpler fractions, called **partial fractions**, that appear on the left side of the equation.*

Decomposition principle (linear case): Assume a fraction is in lowest terms with the numerator having a lower degree than the denominator.

- If the denominator of the fraction contains different linear factors of the form $ax + b$, then for each of these factors the decomposition of the fraction includes a partial fraction of the form $\dfrac{A}{ax+b}$, where A is some real number.

  Example: To decompose $\dfrac{7x-6}{x^2-3x}$ we need to factor the denominator and then provide a partial fraction for each linear factor. Thus,

  $$\frac{7x-6}{x^2-3x} = \frac{7x-6}{x(x-3)} = \frac{A}{x} + \frac{B}{x-3}.$$

- If the denominator of the fraction contains a factor of the form $(ax + b)^m$, where m is a positive integer, then the decomposition of the fraction includes the sum of m partial fractions of the form

  $$\frac{A_1}{ax+b} + \frac{A_2}{(ax+b)^2} + \cdots + \frac{A_m}{(ax+b)^m}$$

  where $A_1, A_2, \ldots, A_m$ are real numbers.

  Example: To decompose $\dfrac{8x+7}{(2x-1)^3}$ we need to provide three partial fractions since $m = 3$. Thus,

  $$\frac{8x+7}{(2x-1)^3} = \frac{A}{2x-1} + \frac{B}{(2x-1)^2} + \frac{C}{(2x-1)^3}$$

KEY EXAMPLE

Find the partial fraction decomposition of $\dfrac{3x^2 - 29x + 80}{x(x-4)^2}$.

Solution: Determine the partial fractions that correspond to each factor of the denominator. For the factor x, there corresponds the partial fraction $\dfrac{A}{x}$, and for the factor $(x-4)^2$ there corresponds the partial fraction sum $\dfrac{B}{x-4} + \dfrac{C}{(x-4)^2}$. Thus,

$$\frac{3x^2 - 29x + 80}{x(x-4)^2} = \frac{A}{x} + \frac{B}{x-4} + \frac{C}{(x-4)^2}.$$

Clear the equation of fractions by multiplying each term by $x(x-4)^2$, the lowest common denominator of the fractions. Thus,

$$3x^2 - 29x + 80 = A(x-4)^2 + Bx(x-4) + Cx.$$

On the right side of the equation perform the indicated multiplications and then collect like terms. The result is

$$3x^2 - 29x + 80 = A(x^2 - 8x + 16) + B(x^2 - 4x) + Cx$$
$$= (A + B)x^2 + (-8A - 4B + C)x + 16A$$

The numerical coefficients of like variable terms on the two sides of the equation, as well as the constant terms, must be equal. Thus,

$$3 = A + B \qquad \leftarrow \textit{Coefficients of } x^2$$
$$-29 = -8A - 4B + C \leftarrow \textit{Coefficients of } x$$
$$80 = 16A \qquad \leftarrow \textit{Constants}$$

Since the solution of this system of linear equations is $A = 5$, $B = -2$, and $C = 3$, the partial fraction decomposition of the given fraction is

$$\frac{3x^2 - 29x + 80}{x(x-4)^2} = \frac{5}{x} + \frac{-2}{x-4} + \frac{3}{(x-4)^2}$$

Denominators with quadratic factors: For each irreducible (non-factorable) quadratic factor of the form $ax^2 + bx + c$ that the denominator of a fraction contains, the decomposition of this fraction requires a partial fraction of the form $\dfrac{Ax + B}{ax^2 + bx + c}$.

Decomposing improper fractions: If the degree of the numerator of the original fraction is not lower than the degree of the denominator, then divide the numerator by the denominator. For instance, before $\dfrac{2x^2 + x - 6}{x^2 - 3x}$ can be decomposed into the sum of partial fractions, we need to divide the numerator by the denominator.

$$
\begin{array}{r}
2 \\
x^2 - 3x \overline{\smash{\big)}\, 2x^2 + x - 6} \\
\underline{-\,2x^2 - 6x} \\
7x - 6
\end{array}
$$

Thus,

$$\frac{2x^2 + x - 6}{x^2 - 3x} = 2 + \frac{7x - 6}{x^2 - 3x}.$$

Since $\dfrac{7x - 6}{x^2 - 3x} = \dfrac{2}{x} + \dfrac{5}{x - 3}$ (see page 203),

$$\frac{2x^2 + x - 6}{x^2 - 3x} = 2 + \frac{2}{x} + \frac{5}{x - 3}.$$

Key 76 Solving nonlinear systems

OVERVIEW   *A **nonlinear system** is a system of equations that contains at least one equation whose graph is not a line. Some nonlinear systems can be solved with the same methods used to solve two-equation linear systems.*

Real solutions of a nonlinear system: The points at which the graphs of the equations of the system intersect, if any, represent real solutions. For some simple types of nonlinear systems these points may be determined by using one of the following methods:

- Graph the equations on the same set of axes and note the points where the graphs intersect. Graphing calculators can be helpful in finding or approximating these points.
- Express one of the equations in terms of a single variable which then can be solved using routine methods. Algebraic methods produce all possible solutions, real and nonreal. If a nonlinear system has no real solutions, then the graphs of its equations do not intersect.

Solving a linear-quadratic system: The accompanying figure shows the possible number of points in which a line and a parabola may intersect. In general, the solution set of a system of one first degree equation and one second degree equation consists of, at most, two ordered pairs.

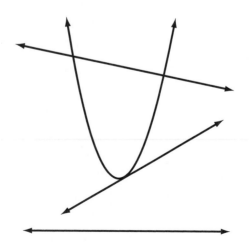

Example 1:   To solve
$$(x - 3)^2 + (y + 2)^2 = 10$$
$$x + y = 3$$

solve the linear equation for x which yields $x = 3 - y$. Substituting this expression in the first equation gives

$$(3 - y - 3)^2 + (y + 2)^2 = 10$$

which simplifies to

$$y^2 + 2y - 3 = 0.$$

The roots of this equation are $y = -3$ and $y = 1$. The corresponding values for x are obtained by substituting each value of y in $x + y = 3$ and then solving for x. Thus, the line $x + y = 3$ intersects the circle $(x - 3)^2 + (y + 2)^2 = 10$ in two points: $(6, -3)$ and $(2, 1)$.

Example 2:   To solve

$$xy + 3y = 7$$
$$2xy + y^2 = 4$$

multiply the first equation by -2 and then add the result to the second equation. This gives $y^2 - 6y = -10$ or in standard form

$$y^2 - 6y + 10 = 0.$$

Using the quadratic formula,

$$y = \frac{6 \pm \sqrt{36 - 4(1)(10)}}{2} = 3 \pm i.$$

The nonreal solutions tell us that the curves defined by the given pair of equations do not intersect.

Solving a system of two quadratic equations: The solution set of a system of two quadratic equations contains, at most, four ordered pairs.

Example 3:   To solve

$$x^2 + y^2 = 13$$
$$xy = 6$$

solve the second equation for y which gives $y = \dfrac{6}{x}$. Substituting this expression in the first equation yields

$$x^2 + \left(\frac{6}{x}\right)^2 = 13$$

or

$$x^2 + \frac{36}{x^2} = 13.$$

Clearing fractions by multiplying each term by x^2 leads to

$$x^4 - 13x^2 + 36 = 0.$$

This equation can be factored as

$$(x^2 - 9)(x^2 - 4) = 0.$$

If $x^2 - 9 = 0$, then $x = -3$ or $x = +3$. If $x^2 - 4 = 0$, then $x = -2$ or $x = +2$. The corresponding values for y are obtained by substituting each value of x in $xy = 6$ and then solving for y. Thus, the rectangular hyperbola $xy = 6$ intersects the circle $x^2 + y^2 = 13$ in four points: $(-3, -2)$, $(3, 2)$, $(-2, -3)$, and $(2, 3)$.

GLOSSARY

Included here are the definitions of many, but not all of the terms used in the Keys. For terms not listed here, please consult the index for page references.

abscissa The x-coordinate of a point in the coordinate plane.

absolute value The absolute value of a number x, denoted by $|x|$, is its distance from zero on the number line. If x is nonnegative, $|x| = x$ and, if x is negative, $|x| = -x$.

additive identity In the set of real numbers, 0 is the additive identity element since the sum of any real number and 0 is that number.

additive inverse In the set of real numbers, the additive inverse of any real number x is its opposite, $-x$, since $x + (-x) = 0$.

antilogarithm The number whose logarithm is given. For example, if $\log 3 = 0.4771$, then 3 is the antilogarithm of 0.4771. In general, if $\log_b x = y$, then x is the antilogarithm of y to the base b.

arithmetic sequence A sequence of numbers or terms in which each term after the first is obtained from the term that comes before it by adding the same constant, called the common difference. The nth term of an arithmetic sequence is $a_n = a_1 + (n-1)d$ where a_1 is the first term and d is the common difference.

arithmetic series The indicated sum of the terms of an arithmetic sequence. The sum of the first n terms of an arithmetic sequence is
$$S_n = \frac{n}{2}(a_1 + a_n).$$

associative law The order in which three real numbers are added or multiplied does not matter.

asymptote A line that a graph approaches, but does not intersect, as x increases or decreases without bound.

axis of symmetry A line that divides a graph so that if the graph is folded on the line the two parts of the graph will exactly coincide.

binomial A polynomial with two unlike terms.

binomial theorem A formula that tells how to expand a binomial of the form $(a + b)^n$ without performing repeated multiplications.

change of base formula The formula
$$\log_a c = \frac{\log_b c}{\log_b a}$$
changes the base of $\log_a c$ from a to b.

characteristic The integer part of a common logarithm.

circle The set of points (x, y) in a plane that are a fixed distance r from a given point (h, k) called the center. Thus, $(x - h)^2 + (y - k)^2 = r^2$.

closure property A set of numbers is closed under an operation if the number that the operation produces is a member of the same set.

coefficient The numerical factor of a monomial.

combination A subset of a set of objects in which order does not matter.

combination formula The combination of n objects taken r at a time, denoted by $_nC_r$, is given by the formula
$$_nC_r = \frac{_nP_r}{r!} = \frac{n!}{r!(n-r)!}.$$

common difference In an arithmetic sequence, the constant that is added to each term after the first in order to obtain the next consecutive term.

common logarithm A logarithm whose base is 10.

common ratio In a geometric sequence, the constant that multiplies each term after the first in order to obtain the next consecutive term.

commutative law The order in which two real numbers are added or multiplied does not matter.

complex fraction A fraction that contains other fractions in its numerator, denominator, or in both the numerator and denominator.

complex number A number that can be written in the form $a + bi$ where a and b are real numbers and $i = \sqrt{-1}$.

composite function The composite function $f \circ g$ is the function that consists of the set of function values $f(g(x))$, provided $g(x)$ is in the domain of function f.

conic section A curve formed when a plane intersects a right circular cone. Circles, ellipses, hyperbolas, and parabolas are conic sections.

conjugate axis The axis of symmetry of a hyperbola that is perpendicular to the transverse axis at its center. The length of the conjugate axis is denoted by $2b$.

conjugate pair The sum and difference of the same two terms, as in $a + \sqrt{b}$ and $a - \sqrt{b}$ or $a + bi$ and $a - bi$.

consistent system A system of linear equations that has a unique solution.

constant A quantity that is fixed in value. In the equation $y = 3x$, x and y are variables and 3 is a constant.

coordinate plane The region formed by a horizontal number line and vertical number line intersecting at their zero points.

counting principle If activity A can occur in m ways and activity B can occur in n ways, then both events can occur in m times n ways.

Cramer's rule A method that uses determinants to solve a system of linear equations.

degree of a monomial The sum of the exponents of its variable factors.

degree of a polynomial The greatest degree of its monomial terms.

Descartes' rule of signs If $P(x)$ is a polynomial function whose terms are arranged in descending powers of x, then the number of positive real zeros of $P(x) = 0$ is given by the number of variations in sign of $P(x)$ or that number reduced by an even number. The number of negative real zeros of $P(x) = 0$ is given by the number of variations in sign of $P(-x)$ or that number reduced by an even number.

determinant of a 2 × 2 matrix If matrix $\mathbf{A} = \begin{bmatrix} a & c \\ b & d \end{bmatrix}$, then the determinant of $\mathbf{A}$, denoted by $|\mathbf{A}|$, is $a \cdot d - b \cdot c$.

dependent variable For a function of the form $y = f(x)$, y is the dependent variable.

dimension The size of a matrix determined by the number of its rows and columns.

direct variation Variable y varies directly as x if $\frac{y}{x} = k$ (or $y = kx$) where k is called the constant of variation.

discriminant For a quadratic equation $ax^2 + bx + c = 0$, the quantity $b^2 - 4ac$.

distance formula The distance between the points (x_1, y_1) and (x_2, y_2) is $\sqrt{(x_2 - x_1)^2 + (y_2 - y_1)^2}$.

distributive law For any real numbers a, b, and c, $a(b + c) = ab + ac$.

domain (relation) The set of all possible first members of the ordered pairs that comprise a relation.

domain (variable) The set of all possible replacements for a variable.

element A member of a set or an entry in a matrix.

ellipse The set of all points P in a plane such that the sum of the distances from P to two fixed points, called foci, is constant.

equilateral hyperbola The hyperbola $xy = k$ $(k \neq 0)$ whose asymptotes are the coordinate axes. Also called a rectangular hyperbola.

equivalent equations Equations that have the same solution set. Thus, $2x = 6$ and $x = 3$ are equivalent equations.

even function Function f is even if $f(-x) = f(x)$ for all x in the domain of f.

exponent In x^n, the number n is the exponent and is the number of times the base x is used as a factor in a product. Thus, $x^3 = x \cdot x \cdot x$.

exponential equation An equation in which the variable appears in an exponent.

exponential function A function of the form $y = b^x$ where b is a positive constant that is not equal to 1.

factor A number or variable that is being multiplied in a product.

factor theorem The binomial $x - r$ is a factor of the polynomial $P(x)$ if $P(r) = 0$.

factoring The process by which a number or polynomial is written as the product of two or more terms.

factoring completely Factoring a number or polynomial into its prime factors.

factorial n Denoted by $n!$ and defined for any positive integer n as the product of consecutive integers from n to 1. Thus, $5! = 5 \cdot 4 \cdot 3 \cdot 2 \cdot 1 = 120$. By definition, $0! = 1$.

field A set of numbers and two operations that are defined on the set such that there exists identity and inverse elements for each operation and the closure, commutative, associative, and distributive properties hold. The set of real numbers forms a field with respect to the operations of addition and multiplication.

focus (foci) A fixed point or points that help to determine the shape of an ellipse, hyperbola, or parabola.

FOIL The rule for multiplying two binomials horizontally.

function A relation in which no two ordered pairs have the same first member and different second members.

fundamental theorem of algebra A polynomial equation has at least one root, real or nonreal. An nth degree polynomial equation has exactly n roots, provided each root is counted as many times as it occurs. The roots may be real or imaginary. If a polynomial equation with real coefficients has imaginary roots, then they occur in conjugate pairs.

geometric sequence A sequence of numbers or terms in which each term after the first is obtained by multiplying the term that comes before it by the same constant, called the common ratio. The nth term of a geometric sequence is $a_n = a_1 r^{n-1}$, where a_1 is the first term and r is the common ratio.

geometric series The indicated sum of the terms of a geometric sequence. The sum of the first n terms of a geometric sequence is

$$S_n = \frac{a_1(1 - r^n)}{1 - r}.$$

greatest common factor (GCF) The GCF of two or more monomials is the monomial with the greatest coefficient and the variable factors of the greatest degree that are common to all the given monomials. The GCF of $8a^2b$ and $20ab^2$ is $4ab$.

greatest integer function The function $f(x) = [x]$ whose function value is the largest integer that is less than or equal to x.

horizontal line test If no horizontal line intersects a graph in more than one point, the graph represents a one-to-one function.

hyperbola The set of all points P in a plane such that the absolute value of the difference of the distances from P to two fixed points, called foci, is constant.

identity An equation that is true for all possible replacements of the variable.

imaginary unit The number denoted by i where $i = \sqrt{-1}$.

imaginary number A number of the form bi where b is a nonzero real number and i is the imaginary unit.

inconsistent system A system of linear equations that has no solution.

independent variable For a function of the form $y = f(x)$, x is the independent variable.

index The number k in the radical expression $\sqrt[k]{x}$ is the root of x to be taken. In a square root radical the index is omitted and is understood to be 2.

inequality A sentence that uses an inequality relation such as $<$ (is less than), $\leq$ (is less than or equal to), $>$ (is greater than), $\geq$ (is greater than or equal to), or $\neq$ (is unequal to).

infinite sequence A sequence whose domain (set of subscripts) is the set of natural numbers.

infinite series The indicated sum of the terms of an infinite sequence.

integer A number from the set $\{\ldots -3, -2, -1, 0, 1, 2, 3, \ldots\}$.

intermediate value theorem If $P(x)$ is a polynomial with real coefficients and $P(a)$ and $P(b)$ have opposite signs, then a real root of $P(x) = 0$ lies between a and b.

inverse of a function The relation obtained by interchanging the first and second members of each ordered pair of the function. The result may or may not be a function.

inverse of a matrix The multiplicative inverse of a square matrix $\mathbf{A}$, if it exists, is the matrix $\mathbf{A}^{-1}$ such that $\mathbf{A}\,\mathbf{A}^{-1} = \mathbf{A}^{-1}\,\mathbf{A} = \mathbf{I}$, where $\mathbf{I}$ is the identity matrix.

inverse variation If y varies inversely as x, then $xy = k$ $\left(\text{or } y = \dfrac{k}{x}\right)$ where k is a nonzero number called the constant of variation.

irrational number A number that cannot be expressed as the quotient of two integers.

leading coefficient For a polynomial in one variable, the number that multiplies the variable with the greatest exponent.

linear equation An equation in which the greatest exponent of any variable is 1.

logarithmic function The inverse of the exponential function.

logarithm of x An exponent that represents the power to which a given base must be raised to produce a positive number x.

major axis The longer of the two axes of symmetry of an ellipse whose endpoints, called vertices, are points on the ellipse. The length of the major axis is denoted by $2a$.

mantissa The decimal part of a common logarithm.

mapping A pairing of each member of a set with exactly one member of another set.

mathematical induction An indirect method of proof. See **Principle of Mathematical Induction**.

matrix A rectangular array of numbers or terms called *elements*. An $m \times n$ matrix has m rows and n columns. The element in the ith row and jth column of matrix $\mathbf{A}$ is named a_{ij}.

matrix addition If $\mathbf{A}$ and $\mathbf{B}$ are $m \times n$ matrices, then $\mathbf{A} + \mathbf{B}$ is an $m \times n$ matrix in which each element is the sum of the elements in the corresponding positions of $\mathbf{A}$ and $\mathbf{B}$.

matrix multiplication If **A** is an $m \times n$ matrix and **B** is an $n \times p$ matrix, then **AB** is an $m \times p$ matrix whose entry in the ith row and jth column is the sum of the products obtained by multiplying the elements in the ith row of matrix **A** by the corresponding elements in the jth column of matrix **B**.

midpoint formula The midpoint of the line segment whose endpoints are (x_1, y_1) and (x_2, y_2) is

$$\left(\frac{x_1 + x_2}{2}, \frac{y_1 + y_2}{2} \right).$$

minor axis The axis of symmetry of an ellipse that is perpendicular to the major axis at the center and whose endpoints are on the ellipse. The length of the minor axis is denoted by $2b$.

monomial A number, variable, or their product.

multiplicative identity In the set of real numbers, 1 is the multiplicative identity element since the product of any real number and 1 is that number.

multiplicative identity matrix For any square matrix **A**, the square matrix **I** that has 1 along its main diagonal and 0 everywhere else since **A I = A**.

multiplicative inverse In the set of real numbers, the multiplicative inverse of a nonzero number is the reciprocal of that number since their product is 1.

natural exponential function The function $f(x) = e^x$ where e is the special irrational number approximately equal to 2.718 . . .

natural logarithmic function The inverse of the natural exponential function, denoted by ln x.

odd function Function f is odd if $f(-x) = -f(x)$ for all x in the domain of f.

one-to-one function A function in which no two ordered pairs of the form (x, y) have the same value of y and different values of x.

ordered pair Two numbers that are written in a definite order.

ordinate The y-coordinate of a point in the coordinate plane.

origin The zero point on a number line.

parabola The set of all points in a plane whose distance to a fixed point, called the focus, is the same as their distance to a given line, called the directrix. The vertex of a parabola is the point at which its axis of symmetry intersects the curve. The directed distance between the vertex and the focus is denoted by p.

perfect square A rational number whose square root is rational.

permutation An ordered arrangement of objects.

point-slope form of a linear equation The equation of a nonvertical line whose slope is m and that contains the point (x_1, y_1) is $y - y_1 = m(x - x_1)$.

polynomial A monomial or the sum or difference of two or more monomials.

polynomial function A function P such that $P(x)$ equals a polynomial in variable x. When the powers of the terms of the polynomial are written in descending order, the polynomial function is in standard form. The leading coefficient is the nonzero number that multiplies the variable with the greatest exponent.

power A number or variable written with an exponent.

prime factorization The factorization of a polynomial into factors each of which are divisible only by itself and 1.

principle of mathematical induction To prove a formula is true for all positive integer values of n, complete these two steps:

(1) Show the formula is valid for the smallest possible value of n, usually $n = 1$

(2) Assume the formula is true for some arbitrary positive integer, say n = k, and then prove the formula is true for the next consecutive integer, $n = k + 1$. This proves that since the formula is true for $n = 1$, it is also true for the next integer, $n = 2$. Since the formula is true for $n = 2$, it is also true for the next integer, $n = 3$, and so forth.

proportion An equation that states that two ratios are equal. In the proportion $\frac{a}{b} = \frac{c}{d}$, the product of the means equals the product of the extremes. Thus, $b \cdot c = a \cdot d$.

quadrant One of four rectangular regions into which the coordinate plane is divided.

quadratic equation An equation that has the form $ax^2 + bx + c = 0$, provided $a \neq 0$.

quadratic formula If $ax^2 + bx + c = 0$, then

$$x = \frac{-b \pm \sqrt{b^2 - 4ac}}{2a} \ (a \neq 0).$$

quadratic polynomial A polynomial whose degree is 2.

radical sign The symbol $\sqrt{\ }$.

radical equation An equation in which the variable appears underneath the radical sign.

radicand The expression that appears underneath a radical sign.

range The set of all possible second members of the ordered pairs that comprise a relation.

ratio A comparison of two numbers by division. The ratio of a to b is the fraction $\frac{a}{b}$, provided $b \neq 0$.

rational function The quotient of two polynomial functions.

rational number A number that can be written in the form $\frac{a}{b}$ where a and b are integers and $b \neq 0$.

rational root theorem If $\frac{p}{q}$ is a rational root in lowest terms of a polynomial equation with integer coefficients, then p is a factor of the constant term of the polynomial and q is a factor of the leading coefficient of the polynomial.

real number A number that is a member of the set that consists of all rational and irrational numbers.

relation A set of ordered pairs.

remainder theorem The remainder obtained by dividing a polynomial $P(x)$ by $x - r$ is $P(r)$.

replacement set The set of values that a variable may have.

root A number that makes an equation a true statement.

scalar multiplication Multiplying a matrix by a constant.

sequence An ordered list of numbers or terms.

sequence function A function whose range is the set of unordered terms of a sequence and whose domain is the set of their position numbers.

Σ (sigma) The Greek letter Σ that indicates a summation of terms.

slope formula The slope of a nonvertical line that contains the points (x_1, y_1) and (x_2, y_2) is $\frac{y_2 - y_1}{x_2 - x_1}$.

solution set The collection of all values from the replacement set of a variable that makes an equation a true statement. Each of these values is a root of the equation.

square matrix A matrix that has the same number of rows as columns.

square root The square root of a nonnegative number n is one of two identical numbers whose product is n. The symbol $\sqrt{x}$ denotes the positive square root x. Thus $\sqrt{9} = 3$ since $3 \times 3 = 9$.

synthetic division A method of dividing a polynomial $P(x)$ by a binomial of the form $x - r$ that uses the detached coefficients of the powers of x.

system of equations A set of equations whose solution, if it exists, is the set of values that make each of the equations true at the same time.

theorem A generalization in mathematics that can be proved.

triangular form of a matrix A matrix in which all the elements below the main diagonal are 0.

trinomial A polynomial with three unlike terms.

transverse axis An axis of symmetry of a hyperbola whose midpoint is the center of the hyperbola and whose endpoints are points on the hyperbola called vertices. The length of the transverse axis is denoted by $2a$.

variation in sign Occurs when two consecutive coefficients of a polynomial in standard form have different signs.

vertex The point at which the axis of symmetry intersects a parabola.

vertical line test If no vertical line intersects a graph in more than one point, the graph represents a function.

vertices (ellipse) The points on the ellipse that are the endpoints of the major axis.

vertices (hyperbola) The points on the hyperbola that are the endpoints of the transverse axis.

zero of $P(x)$ Any value of x such that $P(x) = 0$.

zero product rule If the product of two expressions is 0, then at least one of them must be 0.

INDEX